OTHER BOOKS BY JIM WILSON

Common Birds of Birmingham with Anselm Atkins

Common Birds of Coastal Georgia

Common Birds of the Gulf Coast

Common Birds of Jacksonville

Jim Wilson and Anselm Atkins

Common
Birds

OF GREATER ATLANTA

The University of Georgia Press
Athens and London

Published by the University of Georgia Press
Athens, Georgia 30602
www.ugapress.org
© 1998, 2011 by Anselm Atkins and Jim Wilson
All rights reserved
Designed by Mindy Basinger Hill
Set in 10.5/14 pt Garamond Premier Pro
Printed and bound by Imago
The paper in this book meets the guidelines for
permanence and durability of the Committee on
Production Guidelines for Book Longevity of the
Council on Library Resources.

Printed in China

15 14 13 12 11 P 5 4 3 2 1

Library of Congress Cataloging-in-Publication Data
Wilson, Jim, 1944 Dec. 21–
Common birds of Greater Atlanta / Jim Wilson and
Anselm Atkins.
        p.   cm.
Includes bibliographical references and index.
ISBN-13: 978-0-8203-3825-5 (pbk. : alk. paper)
ISBN-10: 0-8203-3825-7 (pbk. : alk. paper)
1. Birds — Georgia — Atlanta Region.
2. Birds — Georgia — Atlanta Region — Identification.
I. Atkins, Anselm, 1934– II. Title.
QL684.G4W55 2011
598.09758'231 — dc22   2010027880

British Library Cataloging-in-Publication Data available

TO ALL THE PEOPLE

WHO LOVE BIRDS AND SEEK

TO PROTECT THEM

Contents

Very Large Size 17–38 INCHES

Acknowledgments

The authors are grateful for the help they received from many sources. The photographer thanks all of the many people who allowed him to set up and photograph birds in their backyards, including Eileen Breding, Leslie Curran, Ann Brodie Hill, Earl Horn, Karen Pralinsky, Mike Rising, Rusty Trump, Thad and Robin Weed, Victor Williams, and Linda Wilson. Melanie Haire, Frank McCamey, Monteen McCord, and the Chattahoochee Nature Center were particularly invaluable resources, and we greatly appreciate the advice and help of Frank Kiernan, Chang-Kwei Lin, and Jean Torbit at Emory University. Appreciations go to the staff at the University of Georgia Press who kindly helped me through the publishing process of the greater Atlanta book, particularly Nicole Mitchell, who decided this book was worth publishing, Judy Purdy, Mindy Basinger Hill, Jon Davies, and Walton Harris, as well as to my copyeditor Mindy Conner. We also thank Giff Beaton, Barbara Gray, Georgann Schmalz, and Jeff Sewell for checking the photographs and text for errors as we proceeded. Any errors in the final version, however, are the authors' responsibility. Some of the descriptions appeared previously in slightly different form in *Wingbars*, the monthly newsletter of the Atlanta Audubon Society.

Our great appreciation goes to our wives, Kay Wilson and Margaret Kavanaugh, who encouraged and tolerated us and offered numerous helpful suggestions during the years of producing this book.

Introduction

"What bird is that in my yard?" If you live in Atlanta or anywhere else in the Piedmont region, you can probably answer that question by looking through these pages. This book provides descriptions of 61 of the area's most common birds, the ones you are most likely to see, and gives you the means to make a quick identification. Large photographs allow you to match the description with the bird you are watching. Brief, informative text tells you what to observe as you watch your bird and provides other notable facts about the species.

Bird-watching is easy, inexpensive, and fun for adults and children. It is also one of the fastest-growing hobbies in the United States, so don't be surprised if you become hooked. It is indeed a hobby for a lifetime, admitting innumerable levels of interest, commitment, and reward. One reason watching birds is such an enjoyable experience is that anyone can do it, with or without binoculars, and it can be done anywhere—at home, at work, or on vacation. You can bird-watch while working in your garden, eating a meal, puttering in your workshop, or sitting and relaxing. Just look out the window and there they are! If you put up a feeder, you can bring birds even closer to you—close enough, in fact, that you can obtain good photographs with a small camera. And bird behavior is fascinating to observe, whether you're viewing interactions within a particular species or fights between various species over food and territory. Once you have identified a bird, you'll want to know whether it is male or female, young or old, and whether it is a resident, a migrant, or an unusual type. And that's why we wrote this book.

HOW TO USE THIS BOOK

The birds in the book are arranged by size, starting with the tiny ruby-throated hummingbird and ending with the great blue heron. The blue jay and northern cardinal are somewhere near the middle. With size as your guide, you won't waste time searching the wrong part of the book. The sizes given are the measurements of live birds and thus are a little smaller than numbers based on dead, relaxed specimens.

Many of the birds we see in this area don't live here year-round, and some only migrate through. We have excluded these, except for a few very noticeable ones such as the brightly colored indigo bunting and the white-throated sparrow (a common winter ground-feeder). Some birds that live here all the time but are infrequently encountered have been omitted as well; the loggerhead shrike is an example.

Each account describes the characteristics that identify a particular type of bird. The diagram shows the bird parts used in the descriptions. Something is said of the song as well, because although birds don't sing all the time, their song is an important identifying feature if you happen to hear it. Mention is made of the bird's preferred habitat—for example, swamps, thickets, open fields, or high in conifers—and its usual food, its nesting habits, and occasional other interesting facts.

THE NEXT STEP TO ENJOYING BIRDS

Bird-watchers can enrich the experience in many ways that go beyond the thrill of simply watching and identifying birds. A few suggestions follow.

Equipment. Binoculars are indispensable for most bird identification. Start with an inexpensive pair of 7 × 35 binoculars. By the time you want something better, you will know what to buy. As you grow more experienced, you may want a more complete field guide (see the list on pages xviii–xix). To go with your field guides, you may want to obtain some tapes or CDs of birdsongs. Much of your experience with birds will probably be by listening alone.

Web Sites. There are several good Web sites and discussion groups that cover items related to our birds. Among the most useful Web sites are those of the Atlanta Audubon Society (www.atlantaaudubon.org) and the Georgia Ornithological Society (www.gos.org). Georgia Birders Online (GABO) is an email group (reached at GABO-L@listserv.uga.edu) that allows everyone to discuss birds and find out where to go for many species. Hotlines such as Georgia Rare Bird Alert (770-493-8862) can be called for rare bird alerts. Further, many of the Audubon chapters in Georgia have their own Web sites where you can find more about local activities. Go first to the National Audubon Web site (www. nas.org) and from there to the local ones. Finally, some national Web sites, such as the Cornell Lab of Ornithology (http://birds.cornell.edu), allow you to view and hear nearly all the species of birds in North America.

Field Trips. Audubon chapters sponsor group bird walks to see both resident and migratory birds, common and rare. By taking advantage of these free field trips, you can observe and learn hundreds of species of birds. Some of the better bird sites to visit in the greater Atlanta area are marked on the map.

Christmas Bird Count. Every December, the Audubon chapters in the state sponsor holiday bird counts. On this day, groups of bird lovers spend the day identifying and recording every bird they see or hear in specific sites. The

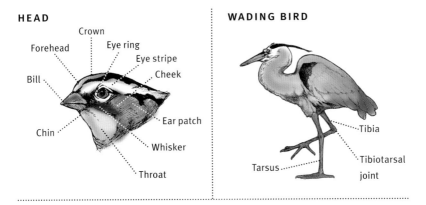

HEAD

Crown
Forehead
Eye ring
Eye stripe
Bill
Cheek
Chin
Ear patch
Whisker
Throat

WADING BIRD

Tibia
Tibiotarsal joint
Tarsus

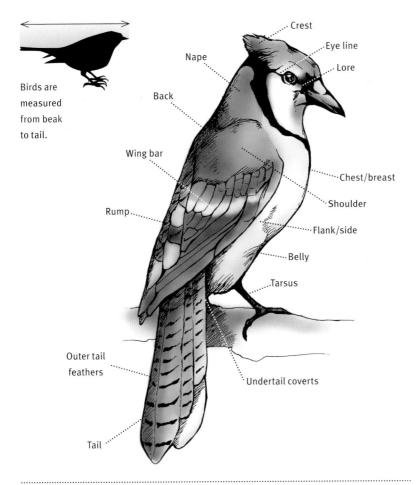

Birds are measured from beak to tail.

Crest
Eye line
Nape
Lore
Back
Wing bar
Chest/breast
Shoulder
Rump
Flank/side
Belly
Tarsus
Outer tail feathers
Undertail coverts
Tail

Bird Parts Mentioned in Species Accounts

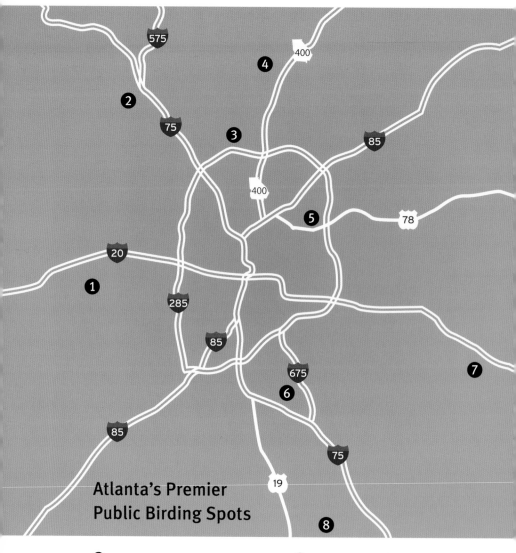

Atlanta's Premier Public Birding Spots

1 Sweetwater Creek State Park
(770-732-5871)

2 Kennesaw Mountain National
Battlefield Park (770-427-4686)

3 Chattahoochee River National
Recreation Area (678-538-1200)

4 Chattahoochee Nature Center
(770-992-2055)

5 Fernbank Science Center
(678-874-7102)

6 W. H. Reynolds Memorial Nature Preserve
(770-603-4188)

7 Panola Mountain State Park
(770-389-7801)

8 Newman Center / E. L. Huie Land
Application Facility (770-603-5606)

For more information on these and other Georgia birding sites, see *Birding Georgia*
by Giff Beaton or visit the Atlanta Audubon Society's Web site at www.atlantaaudubon.org.

atmosphere is fun and casual, but the counts have a serious purpose. The results are sent in to a national headquarters for bird data. The efforts of CBC volunteers have made it possible to compile nationwide statistics on the health or decline of the majority of our species. To participate in a CBC, contact your local Audubon chapter.

Feeder Watch. Cornell University sponsors a nationwide bird count that lets backyard bird-watchers like you make a small contribution to bird research. From November through March, participants set aside two days every other week during which they spend whatever time is convenient observing the birds at their feeders. They also conduct a watch day for migratory birds in May and the Great Backyard Bird Count in February (http://birdsource.org/GBBC/). To sign up, contact the Cornell Lab of Ornithology, 159 Sapsucker Woods Road, Ithaca NY 14850, or visit http://birds.cornell.edu.

Birdathon. Many Audubon Society chapters sponsor an annual birdathon as a money-raising event. Birders find sponsors (often at the workplace) who will donate a given amount per bird species seen. You get to have a wonderful day birding, and Audubon makes money to conserve our birds!

Checklists. A checklist is a list of all the birds that have been observed in a designated place—state, city, refuge, or nature center. The list tells you not only what species might be seen but also the time of year when each occurs and in what numbers (common, rare, etc.). It may also tell you whether the species nests there.

Listing. You may decide to keep a life list—a personal list of all the birds you have identified over the years. More than 700 species of birds can be seen in the United States alone, and about 10,000 in the world. On a smaller scale, you can keep a yard list—species seen or heard in your own yard, which could be well over 100 species! Some enthusiastic birders keep other lists too, such as state or even county lists.

BIRDS IN PERIL

All is not well in the avian world. Humans are hard, vigorous users of the planet and its resources. What we do affects other animals and plants. In fact, our activities are causing the decline or disappearance of many other species. We generally wish no harm to the coinhabitants of our planet, but very often we need or want what they have—and so we take it. Usually what we take is their habitat. And indeed, our appetite for the habitats of plants and other animals has been ravenous. The decline in duck populations, for example, is largely due not to overharvest by duck hunters but to the activities of law-abiding farmers who want to plow their wetlands so that humans can have more food. As the human population grows, the need for food becomes ever larger. Unless we can devise ways to feed ourselves that do not involve taking the animals' space away from them, they will, species by species, go the way of the dodo and the Carolina parakeet.

The abundance—or scarcity—of birds is an indication of how humans are affecting their habitats. As the equatorial rain forests of the world are being leveled at a galloping rate, the migratory birds that winter there dwindle as well. As humans develop the birds' wooded nesting areas in North America, these same birds dwindle all the more. And people who love birds worry about it.

People who love humanity worry about it too, because birds are an indicator of the general health of our environment. When bald eagles and ospreys and peregrine falcons were disappearing because widespread use of DDT was causing their eggs to break and fail, the birds were telling us that too many harmful chemicals were loose in the land. Such chemicals eventually harm humans as well. The indiscriminate release of various chemicals into our surroundings is cause for widespread concern.

Like the miner's canary, our everyday backyard birds can, by their absence, often give the first warning sign of something gone awry with our own habitat. It is for this reason that your local Audubon Society not only encourages your interest in birds, but also reminds you that good conservation habits are essential if we are to continue to live in a world friendly to both birds and humans. As an example, if birds are hitting your windows, use ultraviolet stickers to alert them to this danger; birds have an extra photopigment that allows them to see such stickers as if they were neon signs, making for a quick and easy solution that helps conserve our birds.

OTHER GOOD BIRD BOOKS

As your interest in birds increases, you'll want to consult more detailed guides to bird identification as well as general introductions to the study of birds. We list alphabetically a few that we consider the best.

Erhlich, Paul, David Dobkin, and Darryl Wheye. *The Birder's Handbook: A Field Guide to the Natural History of North American Birds*. New York: Simon and Schuster, 1988. A comprehensive handbook suitable for advanced birders.

Kaufman, Kenn. *Birds of North America*. Boston: Houghton Mifflin, 2000. Good overall book for intermediate and advanced birders.

Parrish, J. W. Jr., G. Beaton, and G. Kennedy. *Birds of Georgia*. Auburn: Lone Pine Publishing International, 2006. Good coverage of all Georgia birds.

Pasquier, Roger F. *Watching Birds: An Introduction to Ornithology*. Boston: Houghton Mifflin, 1980. A good general introduction to the subject.

Peterson, Roger Tory. *A Field Guide to the Birds of Eastern and Central North America*. Boston: Houghton Mifflin, 2002. Lifelike paintings by Peterson; recommended for beginning and intermediate birders. Peterson developed the very useful method of noting "field marks," the particular features that make bird identification quick and certain.

Robbins, Chandler S., Bertel Bruun, and Herbert S. Zim. *Birds of North America*. New York: Golden Press, 1966. Sonograms, or "voice pictures," are included, but the book does not use the "field mark" method.

Schneider, T. M., G. Beaton, T. S. Keyes, and N. A. Klaus. *The Breeding Bird Atlas of Georgia*. Athens: University of Georgia Press, 2010. For birders who want to know general information and nesting ranges for Georgia birds.

Sibley, David Allen. *The Sibley Field Guide to Birds of Eastern North America*. New York: Alfred A. Knopf/Chanticleer Press, 2003. The most comprehensive book on bird features, with drawings of males and females, in breeding and nonbreeding plumages, and juveniles.

Stokes, Donald, and Lillian Stokes. *Stokes Field Guide to Birds: Eastern Region*. Boston: Little, Brown, 1996. Excellent photographs showing different plumages and most field marks. All information is on one page, including up-to-date range map, behavior, and conservation status.

ATLANTA BIRD INFORMATION

Many organizations in Atlanta contribute to the study and conservation of our birds. Concerned citizens who might wish to enlist their help, join, or learn from them can visit the Web sites listed below.

Animal rehabilitation centers: Atlanta Wild Animal Rescue Effort (AWARE), www.awareone.org

Atlanta Audubon Society: www.atlantaaudubon.org

Cornell Lab of Ornithology: www.birds.cornell.edu

Department of Natural Resources, Wildlife Resources Division: www.georgiawildlife.com

Georgia Birders Online (GABO): subscribe via e-mail at GABO-L@listserv.uga.edu

Georgia Ornithological Society: www.gos.org

Georgia Wildlife Federation: www.gwf.org

Humane Society, Atlanta: www.atlantahumane.org

Very Small Size

3–5 INCHES

Ruby-throated Hummingbird *Archilochus colubris*

In the eastern United States, the hummingbird we're most likely to see is the ruby-throated, but only from spring through early fall. In recent years, more and more sightings of western hummers have been reported, but these are winter birds, present after ruby-throats have migrated south. Our summer hummingbirds are almost all ruby-throats.

The male hummingbird's throat feathers, or "gorget," are red, with a black chin above. The back is green; the underparts are white. Females and juveniles lack the throat patch. Because the red on the patch is iridescent, it may appear black under certain lighting conditions.

Ruby-throats are very possessive of their food sources, which are critical for their survival. When several birds are present at a feeder, you'll see them chasing each other away and hear their squeaky, twittery chatter; otherwise just the hum of their rapidly beating wings can be heard. Hummingbirds can assume many acrobatic positions and can even fly backward.

Hummingbirds use their tubular tongue to sip nectar from long-necked flowers, preferring red and orange to other colors. They also take spiders and insects and rob sap from sapsucker holes. Because of their high metabolism and extremely rapid wing beats, hummers need to eat nearly all the time. At night, to save energy, they may fall into a deep torpor that resembles hibernation.

Home feeders are well worth the effort. Fill them with a briefly boiled sugar solution made of four parts water to one part sugar. Change the fluid at least weekly, and clean the feeder at the same time. Do not add red coloring; just have something red on or near the feeder itself.

Female hummingbirds make a tiny nest and feed their young by regurgitation, sticking their long bill right down into the chick's gullet.

The female has white-tipped outer tail feathers.

The female ruby-throat has a white throat — no gorget.

The bright red gorget of the male (above left) appears completely black when the light comes from a different angle (above right).

Brown-headed Nuthatch *Sitta pusilla*

High in the pines you hear a noise that sounds like the squeaky toy in a child's bathtub. You look up and see two or three very small birds darting about in the pine needles. They flit actively from place to place. At that height and against the light of the sky, you can't determine their color; they simply look dark. You can just make out that they have practically no tail. Put those four things together—small size, squeaky toy, pine trees, and short tail—and the mystery is solved: they're brown-headed nuthatches. Our nuthatch is confined to the Southeast, where it can be found throughout the year at lower elevations.

Viewed up close, this nuthatch is a handsome little bird that really does have a brown cap and nape. The cheeks and underparts are white; the back and wings are gray. This nuthatch is even smaller than the very common chickadee, so checking out its characteristics will be difficult unless you entice it close with food.

Nuthatches like black oil sunflower seeds. You may occasionally glimpse one at your feeding station, grabbing a sunflower seed and flying away quickly. Up in the pines, they eat pine seeds and any insects they come across.

For a nest, the nuthatch hollows a hole in a dead pine, often a rather small one. After the young leave the nest, the family forages together for a while in the company of other small birds.

Our brown-headed nuthatch has a western counterpart, the pygmy nuthatch, which is very similar except for its gray-brown cap and peeping song. When the Great Plains were formed, the ancestral species was probably divided into two populations that no longer interbred; each then became a separate species: the brown-headed here, the pygmy there. When you look at them, you're seeing speciation. Those vast, dry, treeless plains served as a divide for many other species too.

The brown cap is clearly visible in both photos.

Note the white breast, gray back, and fairly short tail.

Ruby-crowned Kinglet *Regulus calendula*

Golden-crowned Kinglet *Regulus satrapa*

The **RUBY-CROWNED KINGLET** skitters quickly through the trees searching for insects, often in the company of other small birds. One thing makes the male special: the flaming scarlet crest, which is usually raised only when he gets in a tussle with rival males or is scolding a screech-owl. This kinglet has a light breast with olive-brown or gray wings and back. Its conspicuous white eye ring distinguishes it from the golden-crowned kinglet. The highly visible white wing bars also contrast with the dull bars of the golden-crowned. In the field, look for an eye ring and wing bars. Additional field marks are a short tail and constant wing-flicking. Don't expect to see the red crest.

When you hear the ruby-crowned scold on his wintering grounds in the South, you may think first of a titmouse or wren. His loud song, however, is one you'll never hear on a Christmas bird count.

The **GOLDEN-CROWNED KINGLET** is similar to the ruby-crowned in looks and habits, but with some important differences. First, the golden-crowned has no eye ring, just a solid black eye with a white

The ruby-crowned (top) has a raised red crest when it is excited. The wing bars of the golden-crowned (bottom) have a slight yellowish tint, the dark eyes have a white eye stripe, and the yellow crown is bordered with black (no orange showing).

stripe above it. Second, the golden-crowned kinglet's wing bars are not as distinct as those of its cousin. Third, the broad black stripe at the base of the crown is usually visible even at a distance, as is the gold stripe on the head (the male's has an orange center). The call heard in winter, when the bird is in our range, is a bell-like *tsee*, usually given in threes.

Both species dart actively among the bare twigs of winter trees or search the green boughs of pines for food. They eat berries of honeysuckle and poison ivy, and whatever remains from the bounty of summer. They frequently flick their wings in and out while foraging.

The ruby-crowned has a white eye ring and one white wing bar.

Carolina Chickadee *Poecile carolinensis*

A very common small, gray bird with a black cap, black bib, white cheeks, and white belly is unmistakably our chickadee. They'll come around your house if you have any trees at all, and will be at your feeder in an instant if you put out sunflower seeds. Watch how one eats. It picks through the seeds, casting away those it doesn't want, till it finds just the right one. Then it carries the seed to a nearby branch, holds it between its toes, and pecks at the husk. Soon it flies back for more. When not eating birdseed, the chickadee searches foliage for insects, often hanging upside down from a twig to see what's underneath.

The nest is a hollow excavated in rotten wood or any similar cavity, found or provided, including a nest box. There are usually six eggs, white with reddish speckles. Both parents bring food to the young.

The chickadee sings its name, as many birds do: a rather buzzy *chick-a-dee-dee-dee*. In spring it gives a four-note up-and-down whistled song: *fee-bee-fee-bay*. Chickadees also produce a high squeaking that is hard to describe.

Carolina chickadees like to hang around with tufted titmice, their blood cousins; often you'll see a few of each at the same time. Chickadees will let you get fairly close. Seems like the larger the bird, the more wary it is. Perhaps the chickadee knows it's not big enough to make a mouthful.

This little bird is usually the first to come to a new feeder.

Birds are frequently described as awesome, majestic, pretty, handsome, interesting, weird, or maybe even boring or pesky. But only chickadees can be described as cute. And that is why we can't stop loving them.

Note the tiny bill and striking white cheek
bordered by jet black above and below.

House Wren *Troglodytes aedon*

The house wren is the most nondescript of the wrens, almost completely lacking field marks. Even the light line above the eye that is characteristic of most wrens is too faint to notice. It's just an "all brown" little bird. With a sharp eye on a good day, you might see some light striping along the edge of the wing and tail.

This tiny wren is famed for its aggressiveness. Once his nesting territory is staked out, the male allows no other male house wren near. He even quarrels with his mate over the best way to build the nest (she's extremely picky). He goes around his territory finding suitable holes, putting a few sticks in them, then bringing his choosy mate around for an inspection. If she decides to stay, she tosses out his material and replaces it with her own. The nest may contain strange items. In his account of this species, ornithologist A. C. Bent noted a report of a nest that

The edges of the house wren's wings and tail feathers have slight striping.

contained "52 hairpins, 188 nails, 4 tacks, 13 staples, 10 pins, 11 safety pins, 6 paper clips, 2 hooks, 3 garter fasteners, and a buckle" (*U.S. National Museum Bulletin* 195:113–41). The male is a diligent father. Bent noted another birder who watched a male make 1,217 feeding trips to his family in a single day!

Like many other birds, the house wren likes edges where dense vegetation meets open space. Formerly a winter migrant, it has steadily increased its range southward so that it now nests in and south of Atlanta. It will accept a nest box (make the opening 1¼" high but very wide, so the bird can bring in sticks held horizontally).

The song is an up-and-down gurgling warble that is not at all like the Carolina wren's.

The eye ring is inconspicuous, and the bill is slightly down-curved.

American Goldfinch *Spinus tristis*

The reason the American goldfinch is often called "wild canary" is evident in the male's spring breeding plumage: bright lemon yellow with contrasting black wings, tail, and cap. The female's yellow is very much paler, and she has a dull brownish gray back. In winter, both sexes are brownish gray with just the barest blush of yellow and more muted black. The wing will show one white stripe and a yellow epaulet. Any yellow-colored warbler that looks a little like a goldfinch can be distinguished by its delicate, pointed warbler's bill (warblers eat mainly insects). A goldfinch in flight can often be identified by its undulations and call: *per-chik-o-ree*. The song is long, high, and sweet.

The winter female is gray-brown with very little yellow. The winter male may have a little yellow on the shoulders and face and has no black cap.

Goldfinches are notorious for liking thistles, an expensive seed for feeders. They seem just as happy with sunflower seeds, however, and can be found regularly at the trough alongside other finches. Their thick, stubby bill is good for crushing seeds.

Goldfinches may flock by the dozens or hundreds in trees with seeds and in fields. So devoted is the goldfinch to thistles that it even makes its nest out of their down. Nesting seems to coincide with the maturation of thistles. Farmers, leave those pretty purple thistles in your pastures! Around July, the adults and fledglings may leave their thistle fields and come to your feeders, so make sure you have sunflower seeds as well as some thistle seeds out to accommodate them.

The summer male is bright lemon yellow with black cap and wings and white wing bars.

Indigo Bunting *Passerina cyanea*

INCHES

The names of some birds reflect the varied and subtle colors of nature. Cerulean, lazuli, buff, bay, rose, ferruginous, rufous, ash, ruby, glaucous, chestnut, and clay are only a few examples. The male indigo bunting is truly a rare color—a deep, rich blue. The female has brown wings, a tan head, and a buffy breast with a hint of brown striping. In fall and winter the male resembles his mate except for touches of blue. The male always has a little black on his wings and tail. If you're able to view the indigo's beak, notice its finchlike thickness and its color: dark above, light below. A similar but rarer bird in the countryside is the blue grosbeak (not included here). It is a little larger and is a darker, duller bluish purple in good light, and it has two distinct reddish-brown wing bars.

VERY SMALL   16

City dwellers will have to do a little driving to reach this bird's favored habitat, for it loves the open country. Even there it is present only from spring through fall. The bunting feeds and nests in heavy undergrowth, and you'll seldom discover it on the ground. No matter; the male sings from an exposed perch, frequently in plain view near the top of a tree. You'll find buntings around brushy field edges or on small trees in the field. You can see them on wires and even on the roofs of outbuildings.

The male's song is sweet and distinctive: something like *sweet-sweet, where-where, here-here, see it—see it*. The paired notes are on different pitches, weakening and descending.

The male (top) is all blue except for black touches on his wings and tail. Note the light lower mandible. The female (bottom) has brown plumage, with wings darker than her head. Some light streaking is usually visible on the sides and upper breast. The male indigo is more likely to come to a feeder and perch near your house during its summer visit.

Chipping Sparrow *Spizella passerina*

Like the field sparrow, the chipping sparrow has a light gray, unstreaked breast. Its broad chestnut cap is the best field mark when the chippy is in breeding plumage. Look also for the black stripe through the eye and the white line just above it. In winter, the bird is plainer: the rusty cap is replaced with a dull brown one, the face becomes grayish brown, and the white stripe is drab.

Though chipping sparrows may come to feeders, we usually see them in flocks in the countryside. You can see hundreds feeding in winter corn stubble, hiding in a hedgerow, or feeding on the grass in the front yards of homes. Their flocks are typically larger than those of the field sparrow.

The chippy's conical bill marks it as a seed crusher, but of course it eats insects too. Most perching birds feed their young insects, and these seed eaters are no different.

The chippy's song is a rattle or trill, all on one note. It seems quieter, drier, and faster than the trill of the pine warbler, with which it might be confused. You may need practice to keep from getting the two songs mixed up. Both species are present and singing during our Christmas bird counts, and in spring and early summer.

The Marietta area once had the highest count in the nation for chipping sparrows (thousands), but housing developments have left little food and nesting habitat for these birds. The same is true for many of our declining bird species; 85 species are currently on the Georgia Conservation Priority Bird Species list (see Important Bird Areas on Atlanta Audubon's Web site).

The reddish cap (top) is at its best during the summer breeding season. Also note the black stripe through the eye and the white stripe above it. The rusty cap is replaced with brown during the winter (bottom), and the eye stripe is no longer white.

The lack of a complete eye ring helps to distinguish the chipping from the field sparrow.

Yellow-rumped Warbler *Dendroica coronata*

The yellow-rumped is best noticed and identified by the yellowish splotch of feathers above its tail. Old-time birders lovingly call it the "butter-butt." It is present in Georgia from November through May, with many individuals spending the winter here, and very large numbers migrating through the state in spring and fall. The highest winter concentrations occur along the coast.

In winter plumage, the bird is drab and nondescript, though the yellow rump and a tinge of yellow on the upper breast near the wing remain. But the male in spring breeding plumage is an eyeful with his black eye patch and black pigment on the breast, yellow on the crown, and brighter yellow on the flanks and rump. The western form of this species sports a yellow throat instead of a white one.

The yellow-rump feeds largely on insects, even taking them on the wing like a flycatcher. In winter, it resorts to berries. The song is a fluctuating trill; the call, a *check*. It is often found in flocks in trees or high brush. At those times you can hear its constant little *check* call.

The yellow-rumped warbler is especially interesting because it has two distinct forms, eastern and western, that can be defined as subspecies—an example of evolution in the actual process of dividing one species into two. The two forms can still interbreed when they meet, and thus are still considered a single species. Given enough time, however, they could very well become two different warblers. Of course, this has already happened with our other 40 or so species of warblers, which arose from a single common ancestor. Our eastern form was former- ly called the "myrtle war- bler," and the western form "Audubon's warbler," as if they were indeed two spe- cies, until ornithologists noticed that they were still interbreeding where their ranges overlapped. A very similar cousin, the mag- nolia warbler, has already gone its own way.

The female and the winter male are indistinguishable (above):
dull and brownish, but still with a yellow rump and a little
yellow wash below the bend of the wing.

Male (left) acquiring spring breeding plumage: lots of black,
some white, and bits of yellow.

Carolina Wren *Thryothorus ludovicianus*

The Carolina wren is among our most lovable birds. It is distinguished from other birds by its uptilted tail, and from other wrens seen in Atlanta—such as the house wren and winter wren (not included here)—by its prominent white eye stripe and buffy breast. Larger than other eastern wrens, it is still less than 5 inches long. It ranges over the entire Southeast.

Carolina wrens will nest virtually anywhere: a pair may bring leaves into an old hat hanging in your garage or into a hanging planter—one of their favorite places to raise a family. You might find a nest under the hood of a car, inside your barbecue grill, or even in an unused wheelbarrow! They accept any protected place and seem to like being near human habitation. Pairs raise several broods a year. You can often see little bands of four or five wrens noisily flitting through the brush around your yard.

The Carolina wren lives mainly on insects, which it hunts near or on the ground. It hops briskly over brush piles, tree stumps, and shrubs, searching every cranny. But it will also come to a feeder, where it prefers to eat alone. It seems to like food particles left behind by the other birds.

This wren has an amazing song repertoire boasting at least 125 distinguishable versions. Its bright, cheery songs are much louder than its size would suggest. The notes are in groups of three or four. *Teacher, teacher, teacher* and *tea-kettle, tea-kettle, tea-kettle* are common versions. The wren will also sing triplets of *twinkie, courtesy, piecemeal, trilogy,* and other "words." These aren't random sounds; they're particular songs. It also has a fussy hissing scold, a dry rattling "answering" trill, and a repeated bright rolling trill. Sometimes this wonderful little bird seems responsible for half the bird-song in our yards.

The long, wickedly curved bill; strong white eye stripe; and buffy breast are diagnostic for this species (left). The cocked tail (opposite page) is also typical. Males and females are alike.

Pine Warbler *Dendroica pinus*

It's a dull winter day, you're looking at gray clouds behind a mask of green pines, and suddenly you see a bright spot of yellow. There's only one bird this happy little spot could be: the pine warbler. Just to be sure, you check for field marks: two white wing bars; largish for a warbler, with a stronger-looking beak than usual. Then you confirm: greenish olive above, white toward the back of a yellow belly, a faint yellow eye ring. And just then he lets loose a trilling warble that settles the matter.

Be aware, though, that not all pine warblers are as yellow as the one you see. Females and immatures are duller, and even males can vary. But on your Christmas bird count, even the faintest hint of yellow will be enough to give the pine warbler away.

Pine warblers are in our area year-round. They can be fairly nondescript in the fall, but you can certainly pick them out in winter, when few other warblers are around, and in spring, when the males are in their best plumage. Though pine warblers may sometimes be spotted away from pines, they nest in pines 99 percent of the time and are among the earliest of their group to nest in the spring.

Like other warblers, pine warblers feed mostly on insects. As there are few insects in winter, they may come to your seed feeder on occasion. If you put out a suet feeder, you'll see them all the time.

The trilling warble is similar to but slower than a chipping sparrow's. Just be aware that chippies too can sing from pine trees.

Faint streaking
is sometimes visible
on the sides.

The body is olive above and yellow to pale yellow
below, with two white wing bars and split eye rings.

Field Sparrow *Spizella pusilla*

The field sparrow can be recognized visually by several field marks. The breast is a very plain, pale buff. Some people describe the conical, typically sparrowlike beak as "pinkish," but most see it as bright orange, a color not seen in any of its cousins. The cheeks and crown are streaked with gray and reddish brown. A white eye ring and white wing bars are present as well.

This sparrow is a bird of the countryside. It prefers open areas with small trees, brushy pastures, and abandoned farmland. Unlike the migrant white-throated sparrow, it stays with us year-round; and also unlike the white-throat, it seldom comes to city feeding stations. It sometimes feeds with other birds in winter on the lawns of farmhouses. The flock leaves in parts: a few, then more, then even more; you never guessed so many birds were hidden there on the ground! They scurry to the safety of an overhead tree like leaves blown in a storm.

What food are they looking for? Seeds and small insects. Nesting is usually in thick briers. Young are said to be able to tumble out of the nest prematurely and make it on their own, hidden in the grass.

In late spring and summer, the field sparrow's distinctive breeding song can be heard everywhere in the countryside—in fallow fields, brushy pastures, and grassy yards. The song starts with a few bouncy whistles which grow faster and faster until they become a rising trill. Once you've heard it, you'll know this song forever, and always rejoice in its beauty. Ah, there are field sparrows out there!

Both photos present all the marks needed for identification: pale, buffy breast; pink to orange beak; cheek striped with rust and gray; white wing bars; and white eye ring.

Red-eyed Vireo *Vireo olivaceus*

The red-eyed vireo is one of a group of small, warblerlike birds characterized by a bill that is longer and thicker than that of most warblers and has a little hook at the tip of the upper mandible. Vireos are divided into two clans: those with wing bars and a white or black "mask" around the eyes, and those with no bars or mask, but a white eye stripe instead. The red-eye is in the latter category. Its white eye stripe has a thin black border; above that is a prominent gray cap. For the other characteristics, look for dark olive or greenish back and wings, and white or light underparts. The red eye is not a field mark: it won't be noticeable unless you get a very close look in just the right light.

Like other vireos, the red-eyed feeds almost exclusively on insects. It works its way slowly through the foliage, looking for the caterpillars of moths and butterflies or any other crawly thing. Like all birds, it has keen vision and sees colors well.

This vireo arrives here from its wintering grounds in South America in the spring and either nests here or continues on farther north. In fall, it returns to its winter quarters.

The distinctive nest is suspended between the forks of a branch and secured with spiderweb. The young must all leave the nest at once with the adult; stragglers are left behind. Unfortunately, these young sometimes consist of several

cowbirds, which have parasitized the nest!

The red-eyed vireo has a short, two-phrase song that sounds like *chee-ay*, short pause, *ee-yoo*. It sings this composition over and over, with variations, far longer than most birders want to hear it; but after all, it's not singing for the benefit of birders.

Note the hook at the end of the thick bill and the striping through and above the eye. Most times, as here, the eyes do not appear red.

White-breasted Nuthatch *Sitta carolinensis*

What bird goes down a tree trunk headfirst? That acrobat is the white-breasted nuthatch. It searches for food in all directions: up, down, and sideways. This spry little bird seems never to rest as it gleans spiders and insects from the trunks and large branches of trees.

This nuthatch's breast and throat are white, as its name implies. An area of buffy or rusty pigment on the hind underparts might fool you into thinking you're looking at a RED-breasted nuthatch, but that bird (which is much rarer around here and not included) has a deep rust color on the actual breast and a distinct black eye stripe. Both species, however, have a black (or dark gray, for the female) cap and shoulders, and gray wings.

The female's cap is gray.

This nuthatch abides year-round in our deciduous woodlots and leafy backyards. In winter it comes to feeders, staying only as long as it takes to grab a sunflower seed, which it then opens against the trunk of a tree.

Nesting is in natural cavities and nest boxes. During nest preparation, the male bonds with the female by performing a ceremony at the cavity hole, waving his bill and prancing about. The five eggs are white with a spray of brown on the larger end.

The white-breasted's call announces its presence long before it may be visible: a loud, harsh *quank-quank* or *yank-yank*, given in twos and threes. What is generously called its song is a rapid series of notes on one pitch, amounting almost to a rattle.

As common and easily identifiable as this bird is, it has been surprisingly hard to find on our Marietta Christmas bird counts. Be alert for it, however, almost anywhere in the metro area where there are large hardwoods.

The male (top) has a very dark cap.
Note the short tail in all the images.

Dark-eyed Junco *Junco hyemalis*

The little junco, or "snow-bird," as some bird-watchers call it, is always a welcome winter visitor. Sparrowlike, it seeks seeds on the ground, often under bird feeders. In the countryside, flocks of juncos flit through the tangled undergrowth or dart into the lower limbs of trees.

The dark-eyed junco is so widely variable in its North American range that it was formerly considered to be four different species. Older books list them as "white-winged," "slate-colored," "Oregon," and "gray-headed." All are now considered one species; only the "yellow-eyed" (formerly "Mexican") remains a separate species. Ornithologists change their minds about the classification of species from time to time as new information is gathered and published. Because all four forms can interbreed successfully with each other, they are by definition a single species; when interbreeding stops, new species begin.

Our junco is the "slate-colored" form. It is white on the lower breast and dark gray (slate) above, with the two colors distinctly demarcated at midbreast. Mature males are darker than females, sometimes very dark—almost black—on the head. The seed-crushing bill is pinkish or white, with the thick, conical shape characteristic of sparrows, to which juncos are close kin. The white outer tail feathers are a field mark you can easily see when the bird takes flight.

A few juncos nest in the north Georgia mountains, but most nest farther north. The nest is a cup of bark and grass built right on the ground.

The junco's song—seldom heard in a north Georgia winter—is a sweet trill.

The male (opposite page) is charcoal gray above and white below. The female (below) is lighter gray. The white outer tail feathers of both sexes become easily visible only when they fly.

House Finch *Carpodacus mexicanus*

Purple Finch *Carpodacus purpureus*

The **HOUSE FINCH** lived only in the western United States until 1940, when a few individuals were brought to New York and released. The species spread like wild-fire, eventually reaching Georgia in 1970. The upper breast and crown of the male house finch are tinged with crimson, variable to yellow. The female lacks the red altogether, being instead finely streaked with grayish brown on the underside and no pattern on the head.

You could confuse either sex of the house finch with a **PURPLE FINCH**, which is just common enough in Georgia to make you look twice or maybe three times. The color differences are not reliable, although the male purple finch is generally more raspberry than crimson. The sides of the male house finch have dark streaks, however, while the purple has none. The female purple finch (all brown, also lacking any red) has much heavier streaking than her counterpart, as well as a clear face pattern: a broad, dark band bordered with white.

Purple finches are here only in the winter; they go back north in the spring and nest mostly in Canada. House finches stay here year-round and vie with the feisty house sparrows for nesting sites in the crannies around shopping centers and malls, but they also like patio hanging plants. They sing a bright, many-phrased, and varied warble song.

You'll see them frequently at your

feeder. Like other finches, they sit right there and eat their sunflower seed rather than flying away with it like a titmouse or chickadee. Watch as they crush the seed, letting the two husks fall away. A small flock will feed together rather peacefully on a tray feeder and then suddenly start squabbling with each other.

House finches are here to stay, provided they can develop immunity to a bacterial eye disease that is currently afflicting them.

The brown streaks of the drab female house finch (right) are narrow, and she has no facial pattern; the female purple finch (left) has thick breast streaks and bold white stripes behind and below her eyes.

The male house finch (right) shows brown streaks below his wings, while the male purple finch (left) has none.

House Sparrow *Passer domesticus*

The house sparrow is not a well-liked bird in the United States. Maligned as an exotic import that doesn't belong here, it takes nesting spaces away from our beloved eastern bluebirds and purple martins. It is a scruffy parking lot bird, a dingy city bird, a refuse eater, surely too common for any self-respecting birder to appreciate. Yet those who look with unbiased eyes at the male's striking breeding plumage get a visual treat. The crown is gray, the nape is chestnut, the cheeks are white to gray, and the bib and bill are black. The shoulder is chestnut with one white wing bar. That's a lot of color to crowd into so small a space. The rest of the bird, of course, is just rather sparrowish; ditto for the female and juvenile.

This bird is a weaver finch, not a close relative of our native sparrows. Eight pairs were introduced in Brooklyn in 1850 to help eat cankerworms. That group died out, but later introductions did well enough to spread the species throughout much of the United States.

The breeding male has a black chin,
chestnut wings, and a white wing bar.

You almost never see house sparrows far from human habitation. They nest in any cavity or crevice they can find, often behind some fixture near a building, bringing in quantities of hay and grass, feathers, hair, string, and whatever else they can find. Two broods are usual. They sometimes lay eggs in other birds' nests. These are scrappy and aggressive birds, often stealing from each other and getting into fights. They love dust baths.

Two popular field guides don't even deign to describe the house sparrow's song. Another says it is a long series of monotonous musical chirps. That's true enough, but they're distinctive musical chirps, and the birds sing throughout the year.

House sparrows skittering around on the asphalt of a parking lot as they search for fragments of waste food are a pleasant sight. Look at them closely and think good thoughts. They're among our new "technobirds" and true survivors.

The female and juveniles of both sexes are much plainer.

Song Sparrow *Melospiza melodia*

The song sparrow can be recognized in front view by the tendency of its breast streaks to radiate out from a central area in the middle of the lower neck that often forms a dark spot or splotch. This spot (when it actually exists) can be variable. Sometimes you see it, sometimes you just think you see it. Look for heavy streaking on the breast and a long, rounded tail and gray rump.

Song sparrows love the open country, but they can also be heard singing in town, even near malls and parking lots. You will love the song once you come to recognize it. Think of it as having three parts: some chips, some buzzes, and some trills. With a bit of generosity, you could grant the bird its specific name, *melodia*. Henry David Thoreau translated its song as "Maids! Maids! Maids! Hang up your teakettle-ettle-ettle." Thoreau was a very imaginative person.

Song sparrows do not tend to form large winter flocks as field sparrows and chipping sparrows do, although individuals may appear within such flocks.

You may like this sparrow because of its song or because of the allure of its hypothetical breast spot. But the species is very interesting for other reasons. It ranges widely in the United States and is so variable in characteristics that more than a dozen subspecies can be distinguished. These groups vary mainly in size, but also in bill shape, streaking, and overall coloration. The various populations share their genes back and forth, so further speciation isn't likely right now. But provide some new geographical barriers and a few tens of thousands of years, and this species might erupt into a variety of new kinds of sparrows.

The gray rump is more visible in this view.

This sparrow has a heavy chin mustache coming from the base of the bill, long tail feathers, and plumage that can vary from gray-brown to orange-brown. The heavy, rusty brown to dark brown streaking on the breast usually converges into a spot.

Tufted Titmouse *Baeolophus bicolor*

A prominent gray crest distinguishes the titmouse from all other birds of its size. Notice, too, the gray upper parts contrasting with the white breast. A buffy patch on its side beneath the wing and stark black eyes complete the description.

The titmouse is one of our most common birds around all neighborhoods, winter and summer. Normally it scours leaves for insects, actively flitting from place to place, even hanging upside down from twigs. But it is a reliable feeder bird and particularly fond of black oil sunflower seeds. Watch it take a seed, fly to a nearby twig, hold the seed with its feet, and peck it open. In less than a minute it will be back for another.

The titmouse is a close cousin of the chickadee, and you'll frequently see two or three titmice in company with several chickadees. In breeding season, however, they are, like most other birds, territorial. The titmouse nests in a tree cavity or sometimes in a small birdhouse.

Our bird has a sweet song, much louder than you'd expect for such a small bird. Its *peter peter peter* is a characteristic sound of spring, and it is not above whistling *tweet tweet tweet*. You'll hear squeaky squawks and sucking whistles during food-begging and fighting.

If you are patient, you can gradually train a titmouse to take sunflower seeds from your hand. First, let them learn to eat near you. It helps to have places nearby where the birds can perch while they're deciding whether to brave a possible trap. They may fear your eyes, so squint or close them at first. Be very still. Eventually they may come.

This very common small gray bird is best identified by its crest. If that is not visible, the gray back, lighter belly, and long tail are always apparent. The sexes are alike.

Eastern Bluebird *Sialia sialis*

What species has societies and publications devoted to it and nest boxes put up especially to attract it? The eastern bluebird. Once you see one in the glory of its vibrant blue, you'll be hooked too. The male's back side is an intense sky or cobalt blue. Just as often, though, you'll be glimpsing the rust-colored breast as the bird perches on a wire. Females and juveniles can show much more gray than blue, so be ready to make your judgment by the rusty wash.

The bluebird requires short-grass fields or lawns, in which it goes to the ground to catch its insect prey. For successful breeding, it also needs a very special kind of cavity, not too large or too small, too high or too low. Look for bluebirds on utility lines, fence wires, or the low limbs of trees in open areas. Parents and their young tend to stay together all winter, so any assemblage of three or four small, fat birds on a wire, spaced several feet apart from each other, will usually turn out to be bluebirds.

The bluebird's song is slight and subtle, but sweet—three or four soft gurgling notes.

The male (top) is a glorious blue with a rusty breast, a short tail, and a fat belly. The female (bottom) is less brightly colored than the male but has the same pattern and shape.

Human land-use practices formerly caused a great decline in the bluebird population. More recently, however, thanks to the sustained efforts of many dedicated individuals, bluebirds have made a comeback. Today they are once again a common sight in the open countryside. Making nest boxes and establishing bluebird trails around field edges are very exacting projects. Obtain the proper information before you begin. Bluebirds can be particular, but they will nest in boxes that are appropriately located on a short fence facing a field. Thin metal posts are preferred over trees, perhaps because they deter climbing predators.

In spring, the male brings insects to the female
while she incubates the eggs.

Downy Woodpecker *Picoides pubescens*

Hairy Woodpecker *Picoides villosus*

The **DOWNY** is our smallest woodpecker. If you live in the suburbs, you're sure to see or hear it fairly often. It likes deep woods and dead trees as much as

any woodpecker, but can make do with very little of both. You'll see it around your house, pecking at food on small trees or making a home in a dead branch stub that you wouldn't think fit for any bird. It is petite, busy, and mottled in black and white. The male has a spot of red on the back of his head.

Like most woodpeckers, the downy finds its food by scouring the trunks and branches of trees. Preferring middle heights, it seldom comes near the ground, unlike its cousins the flicker, hairy woodpecker, and pileated woodpecker. It readily comes to suet feeders in winter, and in hard times will even take a fragment from the seed tray.

The song of the downy woodpecker is easy to recognize and often heard. It's a nasal whinny, not too loud, starting high and then descending. There's nothing else like it out there.

The HAIRY WOODPECKER—far less common than the downy—is like its cousin in appearance and habit but noticeably larger (medium sized, not small). The downy is close to the size of a sparrow, while the hairy is more the size of a cardinal.

Look at the bill: thick and long, dwarfing the downy's little stiletto. Its single-note *chip* call is more complex than the downy's. Also, it often feeds low at the bottoms of the trees or on fallen logs—another good reason to let some dead trees lie on the ground.

The hairy (male, top right; female, bottom right) and downy (male, top left; female, bottom left) are difficult to distinguish by plumage, but they differ significantly in body and bill sizes. The hairy is larger, and its bill is as long as the head is wide. The downy's bill length is not nearly as wide as the head: small bird, small beak.

Eastern Phoebe *Sayornis phoebe*

There are many birds in the group called "flycatchers." In fact, this group includes more bird species than any other group in the world. Most flycatchers are infrequently seen and very difficult to identify visually. One that is likely to be seen regularly by bird-watchers in Atlanta and the Piedmont area is the eastern phoebe. Another is the eastern wood-pewee (not included here), but that species is here only in the summer.

The phoebe is easy to identify. First, look for a solitary bird on a low perch. Next, observe its black head, white throat, and breast very faintly washed with yellow. The back is dark gray. Last, but important, watch the tail. The phoebe raises and lowers its tail every few seconds; the pewee doesn't.

The phoebe sits on twigs overlooking open spaces, often near a pond or wetland, and darts out to catch insects on the wing, then returns to its perch. It sits low in a bush or small tree, usually no more than 6 or 8 feet above the ground. These birds enjoy the varied habitat of mixed-use farmland, and they love barbed-wire fences, from which they can dart down into the stubble to capture some unlucky grasshopper.

Phoebes like to nest underneath a natural or artificial structure: on a ledge beneath a rock overhang, perhaps, or under your front porch roof or patio.

We remember the phoebe, and have probably been aware of it all our lives, because it sings its old-fashioned name: a buzzy *fee-bee* or *flee-bee*. The pewee sings its name too: a slurred, whistled *peeeo-weee*.

It has the typical hooked flycatcher bill. Occasionally, you can also see some light yellow wash on the belly. Unfortunately, the photos can't show the wagging tail.

The phoebe's dark head and back contrast with its
light throat, chest, and belly.

White-throated Sparrow *Zonotrichia albicollis*

The white-throated sparrow is with us only during winter and early spring. Everyone who regularly feeds birds on the ground is apt to see it. This sparrow comes in small flocks of four to eight. It's usually found scratching under your feeder early in the morning and late in the evening, the same time most of the cardinals come. It pecks at small seeds awhile, then whisks away to nearby cover. During the rest of the day it stays mostly out of sight, hiding in thick clumps of honeysuckle or privet along streams, or in tangles by brushy fencerows.

The brown, sparrow-sized body and beak and the stark white throat are the best identification characters. If you use binoculars, you'll also see a small yellow patch, or "lore," in front of the eyes that becomes bigger and brighter as spring approaches. The back and wings are brown streaked with black. The breast is pale gray with light streaking. The white-throat is an interesting species because, like the screech-owl, it is dimorphic: it can appear in two forms. One form has a bright black-and-white-striped head pattern; the other has a black-and-tan-striped head pattern. Ornithologists once thought that these represented male and female, respectively, but we now know that either form can be either sex.

When it is hiding in the brush, we recognize the white-throat by his call note, a short, bright, bell-like sound, but a little raspy, as if the bell were cracked. When dogwoods bloom and native trees leaf out in spring, the little white-throat becomes anxious to get back to its breeding grounds up north. At that time you'll begin to hear the male's song, a plaintive whistled theme, sometimes described as "Old Sam Peabody, Peabody, Peabody." You'll hear that for a few days, perhaps a few weeks; then one day, nothing: the bird is gone. You won't see or hear him again until next November.

Adults of both sexes have a clear head pattern, which may be black and white with a bright yellow spot on the lore (top), or brown and tan with a much duller lore spot (bottom). Both photos show the sharply delineated white throat and sparrow bill.

European Starling *Sturnus vulgaris*

The starling is a bird people rave about, one way or the other. Its rather nondescript black feathers become iridescent green under the right lighting conditions during the breeding season. In the winter, the feathers are speckled with white. The strong beak is bright yellow in summer, dark in winter. The wings look very pointed in flight, and when the starling glides, they appear swept back like those of a jet fighter.

The starling is a bird of ill repute both because it's not native to North America and because of some other bad habits. The species was brought from Europe to New York in the late 19th century and spread rapidly across the country. This aggressive bird, equally at home in city and farmland, often steals the nest cavities of popular native species such as bluebirds, red-headed woodpeckers, and purple martins.

Starlings are omnivorous: they eat seeds, insects, fruit, and garbage. They can be seen pecking their way across lawns or scavenging in parking lots. They may flock by themselves or with blackbirds. In mixed flocks, they can easily be distinguished by their stubby tail, pointed wings, and less intense black color.

The starling's song is an interesting and not unpleasant assemblage of squeaks and whistles. The imitative myna bird is one of its kin.

Although it is widely disliked because it competes successfully with certain American birds, the starling is as attractive a bird as many others. Let's restrain our value judgments just long enough to see the starling as a wondrous product of nature.

The male and female are alike, and both wear different summer and winter outfits. The nonbreeding, winter plumage (left) is speckled and the bill is dark; the breeding, summer plumage (opposite page) is almost all black and the bill is yellow. In spring and fall, the plumage can be intermediate between the two.

Brown-headed Cowbird *Molothrus ater*

Bird-watchers hate the brown-headed cowbird; scientists find it fascinating. It makes no nest of its own, and instead lays its eggs in the nest of another bird, often a much smaller warbler or other migrant. Its young take over the nest, usurping the energies of the stepparents, sometimes even ousting the legitimate offspring. It thus wreaks havoc on other species. And yet this unusual breeding strategy, similar to that of the European cuckoo, is worthy of study and perhaps even admiration. If only this fascinating bird were rare and beautiful instead of plentiful and not particularly handsome, how differently might we view it.

The mouse-colored female skulks around at dawn, waiting for her target bird to leave its nest. She darts in, quickly lays a speckled egg, then darts away. Sometimes the returning mother detects the new addition and ejects it; other times not. The cowbird's plan works only half the time, but it costs her nothing more than a few eggs.

The male is black or greenish black with a brown head. Its bill is shorter and more sparrowlike than that of other blackbirds, with whom it often mixes. Its song is a short, squeaky whistle/gurgle; the female gives a soft, rattling chortle.

In the past, the cowbird's range was restricted to the vast prairies of the West and Midwest. Like cattle egrets, cowbirds follow grazing animals to catch the insects they stir up. As agriculture and development opened up the woodlands of the eastern United States, the cowbird moved in. Now it is absent only from the deepest unbroken forests, which are few indeed. Thus, rural development is helping to cause the decline of our forest songbirds.

The male (top) has a two-tone color pattern; the female (bottom) is tan or gray.

The male's iridescence, so common in other black
icterids, makes the bird appear glossy blue.

Cedar Waxwing *Bombycilla cedrorum*

The sleek cedar waxwing, with its sharp crest and black mask, is among our most elegant birds. Brownish amber overall, its tail is tipped with a band of yellow. You may see red waxy tips on some secondary wing feathers if you get a close look at an adult.

Waxwings come to berry bushes in large, bubbly flocks, seeming to pounce on the bush from the sky. They strip it of berries and then move on, often leaving quite a mess beneath.

Flocking is what waxwings do best. Indeed, they are seldom seen singly except during breeding season, which comes late in the year after berries have ripened. Flocks are usually tight and well controlled, wheeling this way and that with precision. In flight they could resemble starlings but are smaller and browner.

Berries are waxwings' staple food, but they also eat insects. Their strangest food, however, is flower petals—apple, for example, and tulip tree. In courting rituals, the male and female pass these back and forth to each other. They nest in the northern Piedmont and the mountains.

The waxwing's song is a high, thin, bell-like trill, like the ringing of tiny chimes. You hear it not as an individual song but as the music of an orchestra.

The subtle breast colors, yellow-tipped tail, black face mask, and head crest are present in both adults and juveniles, but adults (opposite page) also have red-tipped wings (see also p. 53), which the first-year bird (above) lacks.

Rose-breasted Grosbeak *Pheucticus ludovicianus*

For a short time in spring, many people in our region get a rare thrill. Birds with a striking red breast descend on their feeders in little flocks to gorge on sunflower seeds. They hang around for a few short weeks before departing for their breeding grounds farther north. If you crane your neck, you may see the bird high in the new-leafed trees. You're alerted to its presence by its song, which is much like a robin's but sweeter and more energetic. You may also hear its call, a short, squeaky *eek*. Some compare it to the squeak of a tennis shoe on a slick floor.

You'll have no trouble identifying the male red-breasted grosbeak. His triangular red bib draws your eyes, which then admire his coal black head, wings, and tail. The wings have white wing bars, and both the rump and stomach are white. The large, blunt, seed-crushing bill is ivory colored. The female has a dull brownish back and tail, a white breast with brown streaks, a large white eye stripe above a dark cheek patch, and a washed-out golden yellow bill.

The grosbeak's nest is a weak, saucer-shaped affair placed low in an elderberry or small oak. In the southern Appalachians, grosbeaks nest in rhododendrons. The male assists in brooding the eggs.

The grosbeak eats fruit and seeds, cucumber and potato beetles, and the larvae of many other pests.

The migratory rose-breasted grosbeak was added to this book by popular demand. Before the late 1990s, these birds did not often come to feeders. Why they have started coming now is not clear, but everyone wants to know: "What are those gorgeous birds on my feeder?"

The male grosbeak's eye-catching red breast is distinctive even from a distance (top). The female (bottom) lacks the red breast spot and is brown rather than black but has a clear facial pattern formed by white streaks above and below the eye. Both sexes have a chunky body compared with most species.

Red-winged Blackbird *Agelaius phoeniceus*

The black males of this species flaunt a red-orange epaulet at the bend of the wing that is a no-fail field mark. In the breeding season, this epaulet gets even redder and seems to glow, especially when the male spreads his wings in his mating display. Immature males seen in winter flocks have only a pale yellowish bar. Females are entirely different: all brown with a heavily streaked breast. You may mistake them for sparrows if you don't notice the large blackbird bill.

How do you tell a red-wing in a big flock of black-colored birds? First, look for the red wing patch. Compared with the common grackle, the red-wing is smaller and has a smaller bill. In flight, it clearly lacks the grackle's long, wedge-shaped tail. The red-wing does not have the swept-wing jet look of a starling, and it is bigger than a cowbird and has a black head instead of a brown one.

The male (right) is all black except for the bright red shoulder patch, which can be broader than shown here and include orange or yellow below the red. The female (left) has wide streaks on her underparts that are brown and tan rather than black.

In winter, red-wings join other blackbirds to form large flocks that roam the Southeast foraging in fields and yards.

The red-winged blackbird nests in wetlands—beside streams and ponds, and in pastures and weedy fields. In winter, red-wings join other blackbirds to form vast flocks that roam the Southeast foraging in fields and yards. Farmers consider them pests. Flocks can fill the sky like wisps of smoke. A large mixed flock of blackbirds foraging in your yard will feed by "rolling": birds at the rear fly to the front where there is fresh food; in this manner the flock inches forward.

In the breeding season, the aggressive and protective male produces a distinctive *oke-a-lee* song. In overhead flight, red-wings give a quiet *chuck*. The smallish nest is lashed to wetland shrubs and tall grasses. The young can climb and swim even before they can fly. Pairs may produce several broods in a summer.

Eastern Towhee *Pipilo erythrophthalmus*

Our towhee is a ground bird kin to the sparrows. It scratches around among the leaves looking for seeds and insects. The male has a striking black head and back, a white belly, and rufous sides. The female has the same markings, but her upperparts are brown instead of black. The eyes of both are red-orange.

The towhee continually flicks its tail, spreading it open to show the two hidden white feathers. This white-spot flicking is just as much a part of the physical appearance of the bird as any other feature, so be alert for it.

Towhees will come to bird mix spread on bare ground, and sometimes fly up to a feeder. Like brown thrashers, the only time they seek height is when they're singing in the spring. The nest, which is built near the ground, is well concealed and difficult to find.

The towhee is another bird whose call sounds very much like its own name: *toe-weee*. The male's spring song is *drink-your-tea-eee*, but often it says only part of that, and sometimes the *tea* is a whistle. A pair of towhees feeding in the same vicinity will keep in contact with each other by making frequent *chip*s and calls.

The male towhee (above and top left) is most often seen in his familiar ground habitat. The female (opposite bottom) has brown where the male has black.

Red-headed Woodpecker *Melanerpes erythrocephalus*

The red-headed woodpecker is recognized by its black-and-white wing pattern and dark red head. No other woodpecker has a solid red head, yet people often mistakenly give its name to the red-bellied woodpecker, whose crown is the color of an orange-red Crayola. The juvenile red-headed that you see in midsummer is different: its head is gray-brown, and its wings are not quite as black.

Like most other woodpeckers, the red-headed searches trees for insects hiding in or under the bark. Or it may dig for a wood-boring grub. Working hard and taking turns, the male and female can hollow out their nest hole in a matter of days. They prefer to build high in a dead pine, often in a swampy area. In the city, they like large hardwoods growing not too close together, such as in a park or along a street. They would probably use our dead pines if we left a few standing.

In winter, this woodpecker will take sunflower seeds from a tray feeder or go to suet on the side of a tree. On the tray, it is feisty and selfish, often attacking other birds with its spearlike beak.

The red-headed woodpecker sings loudest in spring, when rivalries give rise to raucous singing contests. But it's not much of a song: a rasping whir or a barked nasal squawk. It will drum loudly on a tree to stake out its territory.

This species is declining throughout its range, possibly because starlings drive it from its nesting cavities.

The solid red head and contrasting black-and-white wings show equally well whether the red-headed is on a tree or in flight. The male and female are alike.

Northern Cardinal *Cardinalis cardinalis*

"Redbirds," we call them in the South, because of their nearly unmistakable brilliant red plumage and red-orange beak. So common are they that we take them entirely for granted; yet people travel thousands of miles to see these striking red beauties. The sexes are different. The male is almost completely red except for a grayish wash over the wings. The female has a more brownish or grayish coloration. Juveniles of both sexes look like the female, but the juvenile's beak is brown, not red-orange.

The heavy mandibles show immediately that the cardinal is adapted to crushing seeds, not spearing insects. That stout beak can draw blood from a human's finger. Watch as the bird goes through a pile of sunflower seeds. Seed husks drop out of the sides of the beak as the mandibles grind back and forth to extract the inner meat.

Cardinals will turn up anywhere and have certainly adapted to our suburbs, but they seem to like swampy woods best. In winter, when they aren't being territorial, flocks of 30 or more may take advantage of feeding stations. The range of the cardinal has expanded northward as bird lovers have provided more food during winter.

Cardinals sing mostly in spring, when you'll hear their pleasant *kyeeer, kyeeer* or *purty, purty, purty*. At other times, they utter a single metallic *chip* that you can easily learn to recognize.

Harry Kressing's novel *The Cook* describes redbird hunts with gourmet feasts following. Hunting for the millinery trade once decimated bird populations. Fortunately, it is now against the law to kill or harass any songbird, or even to possess a feather of one.

The prominent crest, orange bill, and black face make this male highly visible as he guards his territory (top). The female (bottom) is brownish gray with some red on the crest, wings, and tail.

Yellow-bellied Sapsucker *Sphyrapicus varius*

Although this name may be an insult when thrown at another person, it is quite appropriate for this woodpecker with yellowish underparts. The yellow-bellied also has the plump body and stout-pointed beak typical of woodpeckers, along with a red patch on a part of its head—in this case, the forehead and forecrown. The male has a red throat patch; the female differs by having a white throat patch. The most distinctive identification feature is a broad white wing patch running from the shoulder almost halfway down the wing that can be seen at a distance.

The yellow-bellied sapsucker migrates the longest distance of any eastern woodpecker. It spends its winters in the Southeast and then heads back to the northern United States and Canada to breed during the spring and summer. There is essentially no overlap in its summer and winter ranges.

The sapsucker may move around considerably from area to area during the winter, or one may pick a tree it particularly likes and peck row upon row of regularly spaced, fairly deep holes in the trunk. It then uses its brushlike tongue to lick out the sap that flows into the holes—thus its name. Ants and other insects that come to the sap are included in the diet as well. The

The male (above) has a red throat; the female (left) has a white throat. Both sexes have white wing patches, which help to identify this species; yellowish underparts; and a red cap.

sapsucker is our only woodpecker whose diet consists of more plant material (sap) than insects and other invertebrates. Unfortunately, damage and even death to the tree can result when the sapsucker makes hundreds of holes in it.

This woodpecker is generally quiet in the winter, but it can produce a cat-like mew or whine (sounding like a downward-slurred *cheeerrrrr*) that is also distinct and identifies it unseen. It tends to be shy and hides behind the tree trunk when humans are around.

Eastern Screech-Owl *Megascops asio*

The eastern screech-owl is a small, chunky bird, no more than 8 inches tall but with a wingspan of nearly 2 feet. Most individuals are rusty reddish brown with white markings on the wings and chest. Some are gray, however, and a rare few are drab brown. Different color phases, or "morphs," like this occur in several species of birds. The causes are not yet clear.

In our area, screech-owls show up wherever there are trees with cavities. Common though these birds are, they conceal themselves well. You're more likely to hear one than to see it. Most of their singing is done in spring and summer. One song is a descending trill or whinny; another is a low, monotonous rattle, very soft. When protecting their young from intruders, the adults make strange barks and popping sounds.

Screech-owls eat birds, small rodents, and large insects. They attack silently on wings muffled by fuzzy feather edges, making no more noise than a butterfly.

Many people render screech-owls more observable by placing nest boxes in their yards. A good box might be 8 inches square and 12 inches tall, with a 3-inch hole near the top. Owls don't want a perch and don't need nest material. Place the box 12 feet above the ground, or as high as your ladder will reach. Owls occupy the boxes as early as December. In late afternoon, or even in midday, the owls may bask in the box opening where you can get a good look at them. The young—two or three—will fledge in May or June. The chicks have gray, downy feathers. Sometimes during the day, songbirds will discover a screech-owl roosting in a thicket or under a canopy of vines and "mob" it, fussing loudly and hopping around excitedly. The owl usually endures this nuisance stoically.

Most eastern screech-owls are rusty brown (top left). The gray phase (bottom left) is less common in the South; the owl shown here has its "ears" flattened. The other two (top left and below) have them raised. Note the yellow irises.

Killdeer *Charadrius vociferus*

The killdeer, a kind of plover, is one of the few shorebirds that we regularly see and hear in Atlanta. Other shorebirds are found inland during migration and may turn up at lucky spots, but killdeer are with us all the time.

The killdeer is easy to identify by its two black bands on a white breast. The back and head are dull brown, but the rump shows a bright rusty color in flight. Its song resembles its common name: *kill-dee*. You can recognize it far away by its piercing piping or whistled notes, *te-te-de-dit, de-dit*.

Open fields and pastures are the killdeer's favorite feeding grounds, but you may also see it by river and lake banks—and, of course, at the beach. Killdeer generally occur singly, searching the grass for worms, crickets, and grasshoppers, although loose feeding flocks of 10–15 are sometimes reported on Christmas bird counts.

This bird nests in gravelly places, easily adapting to driveways and flat rooftops. It makes a "scrape" in which the eggs lie unprotected on bare ground, though they are well camouflaged. The fuzzy young chicks (with just one neck stripe) are precocial—able to run about and feed as soon as they hatch.

The killdeer is perhaps most famous for its "broken wing" behavior. It lures four-legged (and human) predators away from its nest by dragging its wings as if injured and piping plaintively. Just when the cat or fox is sure of an easy meal, the mother bird flies away. Other birds perform this trick too, though none is as famous for it as the killdeer.

The killdeer is often active at evening, making it a "crepuscular" bird. Listen for it after supper on your next camping trip.

Killdeer can also be found in mudflats.

The short, stubby beak and blunt forehead are typical of plovers. In addition to the two black breast bands, the killdeer has a brown back and wings, white underparts, and relatively long legs. Killdeer can be found in fields with short grass as well as in rocky terrains

American Robin *Turdus migratorius*

When robin redbreast sings his best, there can be no doubt that springtime has arrived. Fortunately, robins are with us all year long, hopping on the ground, sometimes singing, nesting at the appointed time. There is no finer song than their springtime *cheerily cheer-up cheerio*, nor a better sight than robins hustling in the snow.

Most people recognize robins: tawny red breast, dark back with black head. The color of the females is a little duller than that of the males. Juveniles sport a speckled breast in summer and have an endearing gawky quality.

Worms are a major diet source for robins.

You can chance upon robins almost anywhere: in fields and yards, in orchards and woods. They usually stay low because they feed on the ground, eating worms and bugs. Don't expect them to be interested in the seed at your feeder.

The nest is made of twigs and straw—and sometimes bits of plastic—bonded with mud or clay, set out on a branch 10–20 feet above the ground. The eggs are a beautiful sky azure. Robins often produce two or more broods per season.

Robins are found in flocks both large and small. A flock observed during a Christmas count was tallied at 1 million birds! But more often you see them in groups of 5–20 birds. When they're flying overhead, you can identify them by their spacing: they seem to like staying 10–15 feet apart.

When you see a robin cock its head at the ground, run a little way, stop, and cock it again, what is it doing? Listening? A scientist who decided to figure it out lay down on the ground to find out what she could hear. She didn't hear much, but she discovered that she could see a lot of insects. And so, probably, does the robin.

The male robin (top), here shown in spring plumage, has a black head and white eye ring as well as the typical orange breast. The female (bottom) looks similar, but her colors are more subdued.

Red-bellied Woodpecker *Melanerpes carolinus*

The red-bellied is such a common woodpecker in our area that you've surely seen or heard one in your backyard. The head of the male is bright orange-red. At your first meeting with this bird you might have thought you were seeing the red-HEADED woodpecker; that bird has a very dark red head, however, with striking black-and-white wings and body. From a distance, our fellow looks grayish because our eye blends its fine black-and-white mottling. Some people call red-bellied woodpeckers "ladder-backs" because of that pattern. A closer look will reveal the mottling and will also show that the whole head isn't orange-red, just the back and top of it. The cheeks are the same light buffy gray as the belly. The female has even less red, just on the back part of her head.

The red-bellied woodpecker willingly eats suet and takes sunflower seeds from a feeder, sometimes fighting for sole possession. It opens seeds by placing them in a crack of tree bark and pecking.

The red-bellied's habits are those of most woodpeckers. It digs its nest hole in a dead tree snag and lays eggs that are round and white, like those of most cavity dwellers. It spends its time on the trunks of trees searching for insects. In spring, it stakes out a territory by drumming on whatever makes a loud noise, including gutters and sides of houses.

This woodpecker's call is a rattle or chortle, intoned in various ways. Distinguishing it from the call of the red-headed takes a little practice. And sometimes you must listen twice to be sure you're not hearing a kingfisher or a great crested flycatcher.

The flight of most woodpeckers is described as "undulating," because they flap in spurts. Between spurts you can see the body arcing like a dart, wings folded. Once you know how woodpeckers fly, you can spot one a mile away.

This woodpecker has red on the crown and nape, not on the cheeks and throat. The red covers the whole top of the male's head (right), but stops at the top of the crown of the female (left). Both sexes show a light red patch on the belly (see title page).

Northern Mockingbird *Mimus polyglottos*

This is the state bird of Florida, and it is common throughout the South. It is "northern" only to South America!

The mockingbird is generally gray with accents of black and white. It has a habit of flexing and extending its wings, showing their large white patches. You see these too when it flies, along with the white edges on the tail. A sleek gray bird with white wing flashes is bound to be a mocker.

Of all the suburban avian singers, only the brown thrasher can rival the mockingbird. The mockingbird's own specific song is long and complicated. But it also mimics other birds and sounds from the surrounding environment.

Sometimes you truly do not know whether you're hearing another bird or the mocker imitating it. And does it sing! For hours, even at night, particularly in the spring. The call, surprisingly, is a harsh *tchack*. The mockingbird will repeat phrases of its song many times, often in groups of threes. Its cousin the brown thrasher

The flashing white wing patches show best when the bird is flying or hunting insects.

sings in doublets. Its other cousin, the darker gray catbird, sings a garbled song of single notes with little repetition. These three species together are our "mimids," so-called for their mimicking abilities.

The nest is within 10 feet of the ground in a tree or bush, and the mockingbirds will attack anything that comes near it.

While some birds get pushed aside by the advance of civilization, the mockingbird thrives on it. The more cell phone towers the better. The more little starter trees on a lawn of manicured grass, better still. House cats? A mere distraction. Learn to appreciate this very special bird, for it seems destined to survive every obstacle we throw at it.

Note the overall gray color with a lighter chest. The body is sleek with a long tail.

Brown Thrasher *Toxostoma rufum*

The Georgia state bird belongs to the same family as that great imitator, the mockingbird. The brown thrasher, however, seldom copies other birds' songs and is considered to be the best avian vocalist in the United States. Its own song, performed in spring from a perch 20–30 feet above the ground, is a series of notes usually sung in pairs. From low in the bushes, especially in the evening and early morning, it will also utter a sharp *chuck* or shushing hiss.

At a distance, a thrasher darting across a shadowy country road could be confused with a female cardinal. But the thrasher is rustier, larger, and has a longer tail, which is often pointed smartly upward when the bird is perched. Notice also the boldly streaked breast; long, curved beak; white wing bars; and pale yellow eyes. Another brown bird with a speckled breast is the wood thrush (not included here), which migrates through our area in spring and fall, often staying the summer; that bird, however, has brown (not yellow) eyes, a white eye ring, and a much shorter tail.

Except in spring, when the males sing from perches, brown thrashers are ground birds. They scratch in the leaves, shuffling and overturning them with gusto in their search for food. They can run quite fast along the ground. Insects and fruit make up most of their diet. They are not common at bird feeders, but may occasionally take some seed from the ground.

The egg is a lovely mottled thing, but it takes luck and skill to find the thrasher's clutch hidden low in the bushes. Of course, if a pair nests in the hedge by your mailbox, it's yours for the viewing.

One of John James Audubon's most dramatic paintings shows four brown thrashers ganging up to protect a nest of eggs from a marauding black rat snake. Nowadays thrashers are much more likely to be destroyed by house cats running loose in the neighborhood than by snakes.

Both sexes have rusty brown wings and back; a streaked breast; yellow eyes; a long, uptilted tail; and a beak meant for business.

Brown thrashers spend a lot of time on the ground
hunting insects in the grass and in leaf piles.

Jays of various kinds are found throughout North America, but the only one in our area is the blue jay. It belongs to the corvid family, which includes crows and magpies. All corvids are smart, noisy, mischievous birds that tend to be scavengers and opportunists. A blue jay will eat anything, from birdseed at a feeder to eggs or babies in a nest.

The blue jay averages 10 inches in length. Its head crest is usually not distinctive. Although we tend to take this common bird for granted, seen with new eyes its subtle colors are remarkable: mauve, blue-gray, blue, and turquoise. The black-and-white markings are striking. The underside is gray, turning to white on the undertail coverts at the base of the tail.

Our jay likes to build its nest at a fork near the trunk of a tree. In early spring, it may construct one or two practice nests before starting on the real one. It likes to build with wet leaves but will use other materials too. It behaves very quietly around its nest, hopping slowly toward it from branch to branch in order to fool predators. It needn't fool you, though: just watch for jays carrying nest material.

The blue jay has a short, sweet, flutelike song. When it isn't happy, it gives a throaty rattle.

Jays are famous for squawking loudly at cats and snakes, but their favorite targets are crows, hawks, and especially owls. A flock of 8 or 10 jays can make a great horned owl really nervous as they fly around and past it, sometimes actu-ally striking the big bird's head and feathers. When you hear mobbing jays giving their most raucous, most frenzied calls, go see if there isn't an owl or hawk nearby.

Blue jays have an unmistakable pattern of blues, blacks, and whites. The blue colors can vary from bright blue to purple-gray depending on the lighting conditions. The blues are the result of refraction of light and not pigmentation.

Mourning Dove *Zenaida macroura*

One bird-watcher in the Southeast who each day listed the birds he saw claimed to have seen a mourning dove every single day of the year. Are they really that common? Not necessarily, but you have surely seen one.

The mourning dove has a tan body with brown and gray wings and gray tail, and here and there a spot of black. Under good lighting conditions you may see some iridescence, especially around the neck and head, which seems small in comparison with the body. The slender beak is slightly down-turned. Strange and special: the wings squeak when the bird flies! The flight is swift and direct, with wings beating quickly and the long, pointed tail trailing behind.

On the ground, the bird walks methodically in the open, pecking at seeds, moving its head back and forth rather like a chicken. It appears to be unable or unwilling to scratch the ground or turn over leaves.

Don't think of the dove as necessarily a bird of peace. On the feeder it will raise its wings in threat and chase off other birds, including its own kind.

Both sexes are light brownish gray with scattered black spots and a long tail that looks very pointed in flight.

The mournful cooing of doves is a characteristic sound of spring and summer. It issues from low, dense shrubs or from perches high in the pines, where the birds build their flat, flimsy nests. Unlike the many species that feed their young on insects, doves nourish their chicks by regurgitation: poking their head right down the chick's gullet to deliver a special food called "crop milk." Adults eat small seeds, even the minute seeds of violets. Their beak clearly lacks the crushing power of finches, cardinals, and sparrows.

Mourning doves are classed as game birds, suitable for sport hunting. This puts them in a class with turkeys, quail, pheasants, and ducks. No wonder the doves we see at our city feeders are very wary: in south Georgia, they are fair game for hunters. In Atlanta, they are preyed on by hawks, owls, and house cats.

The head seems too small for the round body and long neck. The bill is slim and down-curved.

Northern Flicker *Colaptes auratus*

What's that largish brown bird with the white rump that flies up from the lawn when you approach? It's a northern flicker, a member of the woodpecker family. Until recently we called it the "yellow-shafted" flicker, for it shows a flash of bright yellow on the underside of its wings. A similar bird in the West was called the "red-shafted." Zoologists decided these were the same species, so they renamed both birds (along with the gilded flicker) the northern flicker.

Our flicker has a black bib and a red swatch on the back of its head. The belly is strongly spotted with black. The male has a black mustache; the female doesn't. If you see several together on a tree trunk during mating season, check for mustaches to determine each bird's sex. Then you'll know what sort of interaction you're witnessing—love or war.

Flickers are common in Atlanta's suburbs. Elsewhere they prefer the edges of swamps or open areas with just a few trees. Like other woodpeckers, they ordinarily nest in cavities, which they usually drill out themselves.

The flicker's long tongue is good for lapping up ants—both workers and grubs—but it eats other insects too. Berries, including poison ivy, also make up a large portion of the diet.

In the breeding season, you can hear flickers singing *wicker-wicker-wicker* or one of many variations on this. The bird advertises its territory with a short, sharp, screaming cackle. Flickers also have a one-note call, a harsh *screep*, used year-round.

In earlier times, flickers were hunted, for they are good to eat. Nowadays their greatest enemies are owls, hawks, cats, and cars.

The male sports a black mustache. Flickers are frequently seen on the ground looking for insects.

The female does not have a black mustache. Both
males and females have a black breast band, a speckled
breast, and a red dash on the back of the head.

Common Grackle *Quiscalus quiscula*

Aside from crows, the largest blackbird we see around Atlanta is the common grackle. Sometimes called "purple grackle," it may show a bronze, purplish, bluish, or greenish sheen under different lighting conditions. Admire the color, but don't use it as a guide to identification. Instead, look for the long, slightly keeled tail that widens toward the end. The beak is large, and the eyes are pale yellow.

Red-winged blackbirds are considerably smaller than grackles and have a crescent of red or orange or white on their shoulders. Starlings are short and chunky, with hardly any tail at all, and you can sometimes see their pretty speckling. Cowbirds are much smaller than grackles and black with a brown head (males) or mousy gray (females). The common grackle is smaller than its cousin the boat-tailed grackle, which is found on Georgia's coast, and much quieter and better behaved as well.

Grackles often come in huge flocks that descend on your lawn. Listen to their noisy squeaking and creaking as they cover the ground, and watch them turn over leaves in search of nuts and seeds. They especially prefer small acorns such as those of the water oak. In fact, their upper mandible is equipped with a special ridge that helps them crack the nuts open. Grackles strut about as if they think they own the place, but they startle easily. The whole flock may take flight at once just because you raised your window shade or opened the back door.

The grackle's call, according to some observers, is *koguba-leek*.

In winter, grackles cross the sky in huge flocks, often in long, wavy lines. Other blackbirds may be mixed with them, including red-wings, starlings, and cowbirds. All roost together. Forget the saying "birds of a feather flock together"! Look to see what's there.

The grackle may appear bluish black (bottom) or mostly black (top), depending on the lighting conditions, but it always has a long, wide tail and pale eyes.

Rock Pigeon *Columba livia*

Field guides list four species of pigeon (*Columba*) for the United States: band-tailed, red-billed, white-crowned, and rock. The first three are highly restricted in range, but the rock pigeon is found in nearly every town. We know it simply as a "pigeon." All the pigeons in the United States are descended directly from imported domesticated birds.

Pigeons are excellent fliers and are quite beautiful as they wheel and turn in flocks. In some cities, people raise them in rooftop cages and have contests and "wars" with one another's flocks. Charles Darwin is among the most famous historical pigeon fanciers. "Believing that it is always best to study some special group," he wrote in chapter 1 of *Origin of Species*, "I have, after deliberation, taken up domestic pigeons" (1859; Penguin reprint, 1977). There are many different breeds of pigeon, some of them very strange: carrier, short-faced tumbler, runt, barb, pouter, turbit, Jacobin, trumpeter, laugher, and fantail, to name the most important.

Rock pigeons can be a mixture of colors ranging from pure black to pure white, but the most common pattern is dark with an iridescent neck (left) or with a light gray back and wings striped with two dark wing bars (above).

Rock pigeons come in many colors, including white, reddish brown, and black, with gray-blue being the most common; and there are mixtures. But when left to breed on their own, the birds have a tendency to revert to the ancestral form. Let Darwin describe that form: "I crossed some uniformly white fantails with some uniformly black barbs, and they produced mottled brown and black birds; these I again crossed together, and the one grandchild of the pure white fantail and pure black barb was of as beautiful a blue color, with the white rump, double black wing-bar, and barred and white-edged tail-feathers, as any wild rock-pigeon!"

Pigeons inhabit our cities because the stone and concrete buildings are like the rocks on which the ancestral wild species lived. Instead of despising or ignoring our rock pigeons, let's admire their iridescent colors and enjoy their swift flight and aerial acrobatics.

Belted Kingfisher *Megaceryle alcyon*

Our belted kingfisher ranges widely over the entire United States. Lakes, rivers, and streams—even small ones—are its home. The kingfisher is distinctive in both coloration and shape. It has a conspicuous white breast with blue-gray back, wings, and crest, and a breast band of the same color. The female has a second, lower breast band of rusty red. The beak and head are large in relation to the body, and the crest on the head is noticeable even at a distance.

Kingfishers feed by diving headfirst into the water, capturing fish that they swallow whole. They usually hunt from a perch near the bank. Occasionally one will hover like an osprey, and sometimes they will perch on power lines near water. After returning to the perch with a fish, the kingfisher may bash it against the limb quite a few times before swallowing it.

Kingfishers are solitary except when paired for nesting. The nest is a deep hole dug into the side of a clay bank. The sexes take turns digging, sometimes as far into the bank as 15 feet. When one mate comes to relieve the other on the nest, it calls from a nearby perch, then waits for the mate to leave before entering the tunnel.

The call—a thin, dry rattle—is usually given as the bird flies over the water in its characteristic erratic manner. Kingfishers love to fly up and down streams. If the stream is covered with a tree canopy, they fly as if through a tunnel, rattling as they go.

The male (top) has a single breast band. The female (bottom) has a second band that is rusty. Different lighting conditions produce different hues of blue for the back, but the large, shaggy crest is unmistakable.

Wood Duck *Aix sponsa*

The wood duck is generally less common than the mallard—or at least much more secretive—but a far prettier sight once you find it. The male is breathtaking. He has a crest that falls down over the back of his neck like a ponytail. The top of his head is green, his eyes and the base of his bill are red, his cheeks are black with white stripes, his breast is brown with fine, light speckles, his sides are light gold, and his back and tail are black. All that finery is hard to see when he is in flight, though. Instead, look at the shape of the head: the bill will be down-turned. The less colorful female is brown with a wide, distinct, white eye ring. Listen for her alarm call, given in flight: a piercing *whoo-eek*.

Wood ducks almost became extinct in the early twentieth century because people wanted their feathers for finery or desired stuffed specimens for display. They've made a good comeback, due in large part to a program of putting nest boxes for them in ponds and swamps. You've doubtless seen the boxes: wooden structures on poles, usually protected from raccoons by a metal cone. Otherwise, the birds nest in tree

The female's distinct white eye ring is a good field mark.

hollows or large woodpecker holes, in which the female lays 10–12 white eggs.

Wood ducks walk quietly through the forest floor around their swamp, eating acorns. In the water, they dabble on the surface, eating insects and duckweed. They're found in the wildest and most beautiful places: thick swamps with moss draping off the branches. Since they have the ability to take off straight up, without spatting across the water, they can feel safe even in tight places. You may find them on slow-moving creeks and beaver ponds in the city. Usually they see you before you see them, so be prepared for their surprising burst into the air and their eerie cry as they vanish through the treetops.

The beautifully colored male (right) has a high forehead, a crest that looks almost like a backward-facing bill, and much white striping.

Green Heron *Butorides virescens*

Many kinds of birds eat the fish found in our freshwater streams, lakes, and swamps. Kingfishers, ospreys, and bald eagles (not included here) take fish right out of the middle of such waters. Egrets and herons hunt along the shoreline, going in only as deep as their long legs will take them. One short-legged fish eater stakes out the shallowest places of all, and sometimes even hunts from a dry limb just above the water's surface: the green heron.

The deceptive thing about the green heron is its neck. When it is perched or coiled for a strike at its prey, it seems to have no neck at all. That is an illusion, for as it strikes or rises in flight, the neck stretches to a truly heronlike length.

The chest and shoulders, which constitute half the surface area of this heron, are deep, rusty brown. The wings are a dark bronze that, with generosity and imagination, is sometimes called green. In breeding season, the male's legs are bright golden orange; otherwise the heron's legs are pale greenish or ochre.

Green herons nest alone or in small colonies, building their nests anywhere from just a few feet to 20 feet above the water surface or the ground. They have mating rituals and displays as other birds do, including raising of crests and plumes, crouching, and bill snapping.

The large bill attached to the long neck forms an efficient weapon for hunting.

We are seldom privileged to see their courtship behavior, however, and should be quite satisfied just to see one fishing.

The loud call of this heron is distinctive and easy to recognize: a harsh, sharp *skeow* or *kyowk*.

The legs of females and nonbreeding males are pale yellow, but the male's legs become orange during the breeding season.

In its usual shoreline habitat, the green heron's back feathers may look bluish rather than green.

Pileated Woodpecker *Dryocopus pileatus*

The pileated woodpecker is a magnificent reminder of what its recently extinct cousin, the ivory-billed woodpecker, looked like. Early woodsmen, with good reason, dubbed the pileated the "gawd-almighty" bird. Big as a crow and with almost as much black, it glides into your woodlot like a dream or vision. It may peck at a hole it wants for a nest. It may course the trunk of a tree, up and down, looking for ants or other insects. More likely, it will fly right down to the ground to tear apart a rotten log, using its large, chisel-like beak to flake off huge chunks of wood. Grubs and beetles it will take, and centipedes, millipedes, and spiders; but what it really craves are large, black carpenter ants. "Log cock" is one of its local names.

The model for the cartoon character Woody Woodpecker, the pileated gets its name from its conspicuous red crest, or "pileus." Its loud, crazy cackle—a series of 10–15 *kuks*—rings through the woods. The feeding peck (as distinct from territorial drumming) is loud and slow with no rhythm. The bird's flight pattern as it crosses the open spaces of roads and fields is distinctive and easy to recognize. Its song resembles that of the flicker, although the latter's cackle is shriller and faster.

The nest is a deep cavity hollowed out in a sufficiently large dead tree or snag. The vertically oval hole is 3–4 inches in diameter, making it suitable for future use by a screech-owl. The eggs, like those of most cavity nesters, are white and round, but you won't find the shells at the base of the tree; that would alert predators to the presence of the nest.

The pileated woodpecker is fairly common in our towns, and relatively unwary. Feeding with little regard for human intruders, it often allows a close approach. But whether you view it with or without binoculars, the sight is always spectacular.

Both sexes have solid black wings, tail, and body with a bright red crest. The male (right) has a red mustache behind and below the bill, while the female (left) has a black mustache.

Cooper's Hawk *Accipiter cooperii*

The Cooper's hawk is not as well known to most people as the red-tailed hawk. It is smaller, with shorter wings but a longer tail. The breast of mature birds is reddish with horizontal barring, and the back is slate gray. Immatures have a light breast with vertical streaking and a brown back. A person is lucky to see a Cooper's hawk in a day's birding in Georgia, luckier still to hear its shriek or its *cack-cack-cack-cack*.

The breast has horizontal banding (top). The female's back can be browner, but her long, broad, white-striped tail and breast banding are still prominent (bottom).

Hawks are a confusing group at first, but with persistence you can straighten out the ones likely to be seen in your area. Even after that, though, you'll probably have trouble distinguishing the Cooper's from its close cousin and fellow accipiter, the sharp-shinned hawk (not included here). The sharpy breeds north of here in the summer, but in the winter it is as common as the Cooper's. The little sharpy looks very much like the Cooper's, but is smaller and has a less rounded tail.

The Cooper's hawk is a bird of woods and streamside groves. It lives generally by eating other birds (as does the sharpy), which it surprises with swift and agile pursuit within the close confines of the trees. Like other accipiters, the Cooper's seldom soars, preferring to flap and glide. It nests in conifers.

Our bird lives year-round over much of the United States, though it is infrequently seen. It might appear in a suburban backyard only fleetingly, though if it finds unwary birds at a feeder, it may hang around for a while.

The adult male has a bluish gray back (from which it receives the local name "blue darter") and a rounded, banded tail with a white border.

Red-shouldered Hawk *Buteo lineatus*

The red-shouldered hawk is a buteo, member of a group of soaring hawks with short wings and stout bodies. However, it has longish wings and a longish tail for a buteo. It flies somewhat like an accipiter, with flaps and glides, though it may also soar like a red-tailed hawk. As soon as you hear its *kee-yar, kee-yar* or *kyear, kyear, kyear*, you know instantly what it is. Unlike the raspy whistle of its cousin the red-tail, you can hear the call of this hawk for miles. Blue jays frequently imitate it, but the hawk's cry is louder and stronger than the blue jay's—and harsher and more fearsome if you are close to it.

Perched, the mature bird shows a reddish belly or a back with red-brown shoulder patches. In flight, the black tail barred with white is a dead giveaway.

Whereas the red-tail can be found just about anywhere, the red-shouldered prefers areas near streams, swamps, and moist woods. Its diet is mainly aquatic: frogs, turtles, and snakes, varied at times with rodents, rabbits, robins, screech-owls, crows, wasps, and grasshoppers. Though colloquially termed "hen hawk," it seldom takes poultry.

Red-shouldered hawks stay paired for many years, possibly for

The mature bird shows a reddish belly and narrow white tail stripes.

life; the record is 26 years. The nest is built of twigs, bark, leaves, and softer things, and generally resembles that of the red-tailed hawk but is smaller. Pairs may return to the same nesting site season after season.

The reddish shoulders
and horizontal barring
on the breast are reliable
field marks (opposite).

Mallard *Anas platyrhynchos*

The mallard is one of our most common ducks and certainly is the one most people can name. It thrives in city and farm ponds. Individuals may stay around because their wings have been clipped, but most stay because they know where free food is!

The drakes, or males, of all our duck species have distinctive plumage. The mallard drake is easily recognized by his large yellow bill and green head with a white neck ring. You may also spot his recurved tail, from which we get the term "duck-tail" (as in the haircut). After the mating season, the drake loses his bright colors and assumes a duller "eclipse" plumage. Female ducks, or hens, are usually drab brown, and the various species are sometimes difficult to distinguish from one another. The mallard hen is a brownish bird with an orange bill marked with black, a dark eye stripe, and a blue wing patch, or "speculum," that is present in both sexes but not always visible.

Ducks can usefully be divided into several groups. The mallard is a dabbler, or puddle duck, meaning that it feeds on a shallow pond or lake bottom by tipping its rear end skyward. It may also walk and feed on land. The pochards, or diving ducks, have legs set far back to help them swim and feed underwater, though this makes it awkward for them to walk on land.

The mallard offers an example of how bird-watching and conservation go hand in hand. Most of our ducks require freshwater ponds and wetlands for their survival. Americans have converted much of our wetland acreage into land for agriculture and living space. Such a practice, if continued, could lower the populations of ducks and other birds, taking some to extinction. For this reason, many birders become avid conservationists, supporting public policies that protect habitats for wildlife. Mallards have benefited greatly from their efforts.

The drake (top) has a yellow bill, green head, white neck stripe, and a curled tail with black feathers that are clearly visible. The hen (bottom) has an orange bill with a dark top and a dark eye stripe. Both sexes have a blue speculum, although the male's is hidden here.

American Crow *Corvus brachyrhynchos*

When crows wheel and circle overhead, you cannot ignore them. Their flight is a thing of beauty. They are handsome birds, too: solid black from beak to feet, with a big, thick bill. Crows are smarter than many birds: they can count up to three or four, and they're notoriously wary.

Crows are two or three times larger than any of our grackles or blackbirds, but much smaller than the vultures that soar in the sky, and somewhat smaller than the ravens that inhabit remote regions of our mountains. Their size varies substantially. The only other bird you are likely to confuse with our American crow is the fish crow, which lives mainly on the coast, although small numbers have recently appeared around Atlanta. Go by sound. The common crow says *caw, caw* or *awk awk awk awk*. The cry of the fish crow is a very nasal *car-ar*. Alas, the summer juveniles of common crows can sound a lot like that! But if your bird says *caw*, then for sure it's an American crow. Novice birders sometimes confuse crows with hawks. Crows are solid black, their beaks are straight and not hooked, and they tend to sit on the very tops of trees instead of a little way down.

Crows usually roost and nest in isolated places. They sometimes gather in flocks of hundreds, but it's more common to see them in groups of four or five.

Crows have many ways of making a living. As omnivores, they'll eat grain, carrion, or anything else edible that opportunity presents them (peanuts in your backyard, eggs in a nest). That bodes well for this adaptable species.

Great horned owls and red-tailed hawks sometimes use abandoned crows' nests (stick platforms with deep centers). Crows help bird-watchers by pointing out owls and hawks, which they mob in large numbers (5–25). They carry on for 10 or 15 minutes, cawing stridently and making passes at the owl. Keep your ears cocked for a mobbing, and you'll find owls and hawks there.

Crows are two or three times larger than any of our grackles or blackbirds.

No other solid black bird in our area is as large as
the crow and has eyes as dark as its body.

Barred Owl *Strix varia*

The barred owl is a large bird, although slightly smaller than the great horned owl. It haunts swamps and bottoms, eating small animals typical of wetlands. In size and habitat it stands to the great horned owl as the red-shouldered hawk does to the red-tailed. From the front, observe the strong vertical dark-and-white barring (hence the name) on the breast. See the round head (no "horns") and the black (not yellow) eyes.

The most exciting thing about the barred owl is its song. Known familiarly as the "eight-hooter," this owl's call is often described as "Who cooks for you, who cooks for you-all?" But it has many variations on this basic song—all of them rather loud, some of them downright bizarre. It may sing just the first half of its song, or it may add or subtract a note along the way. Often it gives a single long, descending, guttural *hoo-aww*. When it sings in duet with its mate, it barks and caws. Owl-prowlers sometimes ask: Was that an owl or a dog? By contrast, the great horned owl's song is soft, usually consisting of four or five notes that sound like "who, who, who, who."

Barred owls frequently nest in cavities and will also use an old hawk nest or a properly designed nest box. Breeding and nesting begin in late fall or early winter, enabling the young to have a long growing season.

The barred owl hunts actively at night, in early morning, and even in midday. It may allow humans to approach fairly close, and can be lured into view by the proper calls (birders generally use recorded calls for this purpose).

The dark brown vertical barring on the breast gives this bird its name.

The dark eyes and lack of "horns" easily distinguish the barred owl from the great horned. The owl can turn its head to face backward.

Red-tailed Hawk *Buteo jamaicensis*

The red-tailed hawk is common and widespread over the entire United States, soaring and hunting widely in open country. This buteo is quite successful as a bird of prey, as evidenced by its healthy numbers and many subspecies. As long as humans don't poison the rodents it eats, this hawk will get along just fine.

You may see the red-tail perched along the interstate, white breast shining in the sun. Or you may see it gliding into the grass of the median for a kill. It is dark brown above with white underparts and usually a dark brown breast band. Both sexes have the reddish tail. The immature, however, has a drab tail banded dark and light with gray or brown.

The red-tail finds a tall tree in which to build a flat nest of twigs and sticks that is more than 2 feet in diameter. The pair continues to bring new green sprigs as long as the eggs are incubating. During this time the male also brings food to his mate. Nests may be reused.

It takes some practice to tell a soaring red-tail from a vulture. They're about the same size and soar in the same way, sometimes even together. But a vulture is black, and a red-tail will appear brown. The turkey vulture holds its wings in

a dihedral—a shallow u or v—shape; the hawk holds its wings flat. If you can view the undertail against the sunlight, you'll see the hawk's reddish tail feathers showing through as a faint pink: definite identification.

The red-tail may hunt from perches near open fields or from the air. It may perch high—two-thirds of the way up a tree—or as low as a fence post. It wants small rodents such as rats, mice, rabbits, and squirrels but will take snakes and eat roadkill. On rare occasions it takes a bird. Chickens? No. This is not a chicken hawk. Help discourage farmers from shooting this beneficial bird: it eats the rats and squirrels that steal his corn!

The red-tail's call is a rasping, one-note whistle that sounds almost like steam escaping.

Immature birds have stripes on their tail (barely visible here).

The heavy body and short tail are typical buteo traits. The undertail faintly washed with red, dark hooded head, and dark breast band are reliable field marks. Our hawk's tail color is about that of a robin's breast and is easily seen against a blue sky (above).

Great Horned Owl *Bubo virginianus*

The great horned owl generally calls with five soft hoots: *hoo-huh-hoooo, hoo-hoo*. At night you can easily tell it from any other owl by this song alone. This is the largest owl in our area. It is named for its "horns," which are actually nothing more than tufts of feathers. Look for yellow eyes (the nonhorned barred owl has black eyes) and a prominent large, white bib under the throat. The light breast is barred with horizontal striping (the barred owl has vertical striping). The back is mottled brown and gray. There are 12 subspecies north of Mexico, each with its own slight color variations.

The great horned uses its wickedly sharp talons to capture anything that moves from dusk to dawn: squirrel, skunk, bird, rat, and snake. As is true of many birds of prey, the female is larger than the male. Her size gives her a little more control and protection during the mating process, and makes her a proficient hunter of larger prey.

The great horned roosts by day in deep shade, where it is often harder to see than you might expect, considering its size. The hiding place is sometimes disclosed by mobbing crows, whose frenzied chorus of caws alerts the observer to its presence.

Great horned owls begin breeding in late fall and lay eggs in early winter, often in an abandoned crow or hawk nest. The chicks, one or two, will hatch out during early spring. These birds get an early start because they need a long time to grow in order to survive the coming winter on their own. The chicks are fuzzy gray and surprisingly large. Their food-begging call, given about every half-minute, is a loud, rasping *screep*.

The great horned owl manages to survive in the metro area perhaps because so many large city trees escape the logger's saw and because we have so many squirrels and other small mammals.

The "horns" may be relaxed and flopping to the sides or sticking up as if to make the bird look fiercer. The yellow eyes with black pupils, white bib, and horizontal barring on the breast are also field marks. The reddish brown facial disk is worth observing too.

Black Vulture *Coragyps atratus*

Turkey Vulture *Cathartes aura*

Turkey vultures and black vultures often soar and feed together during the day, though usually turkey vultures are more prevalent. Make sure to inspect any group of vultures to see if some of them look different from the others. The black vulture has an all-gray head, not the blood-red head of the turkey vulture. The black vulture also has whitish to light gray feathers only on the tips of its wings, while the turkey vulture has light gray from the wingtips to the body along the trailing edge. The latter characteristics are easily noted when the bird is soaring overhead. Farther away, notice that the black vulture flies with its wings straight out, while the turkey vulture's wings meet its body in a pronounced U or V. This V shape, which can be seen from miles away as the TV soars through the skies looking for a meal, also distinguishes it from the hawks—usually the red-tailed hawk—that soar in our southern skies. The tail of the black vulture is extremely short, so short that wings and tail seem to blend into one. Some people call the black vulture "short-tail."

Both species nest on the ground in some dark, secluded spot, and both feast on carrion. We see both species tugging at carcasses beside interstates and other heavily traveled roads. Black vultures do not have a very good sense of smell and locate their food by sight alone or by following turkey vultures, which do have a good olfactory system.

Vultures rarely vocalize, so listening is not a useful way to distinguish them.

The turkey vulture has a red head (above top) and light gray
along the entire rear half of its underwings (above bottom).
The tail is wedge-shaped.

The black vulture (opposite left) has a gray head, and
only the tips of its underwings (opposite right) are light.
Note the short tail.

Canada Goose *Branta canadensis*

In recent years, these stately birds have forsaken their migratory nature and taken up permanent residence in metro Atlanta ponds and lakes. Some residents consider them a great nuisance; and such, indeed, they may have become.

There is no mistaking a Canada goose. No other goose has the long black neck with the stark white underchin. No other is so loved by bread throwers at city ponds and parks, or so hated by lakeside homeowners who must deal with their excrement. On the other hand, they're said to be quite tasty.

Canada geese mate for life. They are wary, strong, and protective of their young. Do not come between the parents and their goslings! An enraged goose is a fearsome bird.

The distinctive *ka-runk, ha-lunk* or *honk-a-lonk* calls of migrating geese are familiar sounds in the fall as their v formations pass overhead going north to south. Bird-watchers should be aware, however, that the migrant most often seen and heard over the Atlanta area is not the Canada goose but the sandhill crane (not included here), whose call is a gurgling trill, not a honk. Sandhills regularly fly over Atlanta en route between their wintering grounds in the Oke-fenokee Swamp of southern Georgia and northern Florida and their breeding grounds in the Northwest. Let "sandhill crane" be your first thought when you catch sight of a large v of migrating birds.

Ornithologists divide the Canada goose into subspecies, each with its own distinctive characteristics. Size, breast color, and neck ring all bespeak variations in the genome passed from generation to generation. Wild populations differ greatly. How else could they ever have evolved into the multifarious forms we see around us?

The male and female look the same.

The black neck and head and white chin strap and cheeks are unmistakable.

Great Blue Heron *Ardea herodias*

The great blue heron is Atlanta's largest waterbird. It has a 6-foot wingspan and stands 3 feet tall. The feathers are steely blue-gray, and the shoulders have a black patch. This heron is often seen standing at the edge of pond or stream with its long neck either held straight or curved into a graceful s. It flies with deep, slow strokes, legs extended and neck folded into a tight curve. Cranes and geese fly with their necks straight out.

The great blue could be confused with the yellow-crowned night heron (not included here), which is smaller, shorter, and less common. The night heron has a short black bill, whereas the great blue heron has a long yellow one. If you are seeing it from a good distance, the only other possible bird you might think it resembled would be a wood stork or sandhill crane (not included here), neither of which would be seen on the ground around our area.

With its neck extended, this bird is more than 3 feet tall, giving it an excellent view of fish. That size makes the great blue unmistakable even from a distance! The great blue is usually seen at the water's edge.

Ranging widely over most of the United States, the great blue feeds by spearing fish with its bill, tossing them in the air, and catching them headfirst for swallowing. Given the opportunity, it stakes out backyard fishponds and feasts on a gourmet meal of koi or goldfish. When alarmed, it flies off with a harsh grunt; otherwise it is silent. Look for great blues along the banks of rivers, in ponds of all sizes, and along marsh edges.

The nest is built of sticks in a wetland setting. In the last century and the early part of this one, the feathers of herons and egrets were prized as decorations for ladies' hats. All of our herons suffered population declines as a result of excessive hunting for their plumage, which in weight was worth more than gold.

The great blue curls its neck during flight and while
preparing to spear fish with its daggerlike bill.

Checklist of Common Birds of Greater Atlanta

	STATUS	SPRING	SUMMER	FALL	WINTER
American Crow	PR	A	A	A	A
American Goldfinch	PR	C	C	C	C
American Robin	PR	A	C	A	A
Barred Owl	PR	U	U	U	U
Belted Kingfisher	PR	C	C	C	C
Black Vulture	PR	U	U	U	U
Blue Jay	PR	A	A	A	A
Brown-headed Cowbird	PR	C	C	C	C
Brown-headed Nuthatch	PR	C	C	C	C
Brown Thrasher	PR	C	C	C	C
Canada Goose	PR	C	C	C	C
Carolina Chickadee	PR	C	C	C	C
Carolina Wren	PR	C	C	C	C
Cedar Waxwing	WR	C	—	U	A
Chipping Sparrow	PR	C	C	C	C
Common Grackle	PR	A	A	A	A
Cooper's Hawk	PR	U	U	U	U
Dark-eyed Junco	WR	C	—	C	A
Downy Woodpecker	PR	C	C	C	C
Eastern Bluebird	PR	C	C	C	C
Eastern Phoebe	PR	C	C	C	C
Eastern Screech-Owl	PR	C	C	C	C
Eastern Towhee	PR	C	C	C	C
European Starling	PR	A	A	A	A
Field Sparrow	PR	C	C	C	C
Golden-crowned Kinglet	WR	C	—	C	C
Great Blue Heron	PR	C	C	C	C
Great Horned Owl	PR	C	C	C	C
Green Heron	SR	C	C	C	R
Hairy Woodpecker	PR	U	U	U	U
House Finch	PR	C	C	C	C

	STATUS	SPRING	SUMMER	FALL	WINTER
House Sparrow	PR	A	A	A	A
House Wren	PR	C	C	C	U
Indigo Bunting	SR	C	C	C	—
Killdeer	PR	C	C	C	C
Mallard	PR	C	C	C	C
Mourning Dove	PR	A	A	A	A
Northern Cardinal	PR	A	A	A	A
Northern Flicker	PR	C	C	C	C
Northern Mockingbird	PR	A	A	A	A
Pileated Woodpecker	PR	C	C	C	C
Pine Warbler	PR	C	C	C	C
Purple Finch	WR	C	—	C	C
Red-bellied Woodpecker	PR	C	C	C	C
Red-eyed Vireo	SR	C	C	C	—
Red-headed Woodpecker	PR	U	U	U	U
Red-shouldered Hawk	PR	C	C	C	C
Red-tailed Hawk	PR	C	—	C	C
Red-winged Blackbird	PR	A	A	A	A
Rock Pigeon	PR	A	A	A	A
Rose-breasted Grosbeak	M	C	—	C	—
Ruby-crowned Kinglet	WR	C	—	C	C
Ruby-throated Hummingbird	SR	C	C	C	—
Song Sparrow	PR	C	C	C	C
Tufted Titmouse	PR	C	C	C	C
Turkey Vulture	PR	C	C	C	C
White-breasted Nuthatch	PR	C	C	C	C
White-throated Sparrow	WR	A	—	A	A
Wood Duck	PR	C	C	C	C
Yellow-bellied Sapsucker	PR	C	—	C	C
Yellow-rumped Warbler	WR	C	—	C	C

Image Credits

I provided most of the images, but for those I did not, I thank the following sources:

p. xv: Alison Cundiff
pp. xx-1: Morguefile.com
pp. 2-3: Stefan Ekernas / Dreamstime.com
p. 8 (top): Todd Schneider
p. 16: Tony Campbell / Dreamstime.com
p. 17 (bottom): Howard Cheek / Dreamstime.com
p. 47 (bottom): Giff Beaton
p. 63: Rinusbaak / Dreamstime.com
p. 98: stubblefieldphoto / Dreamstime.com
p. 105: Hokoar / Dreamstime.com
p. 117 (top): Jerry Amerson

Index